DATE			

Radar Homing Guidance for Tactical Missiles

Radar Homing Guidance for Tactical Missiles

D.A. James

formerly Principal Lecturer in Electronics
Royal Military College of Science
Shrivenham

A HALSTED PRESS BOOK

JOHN WILEY & SONS
New York

First published 1986 by

MACMILLAN EDUCATION LTD
London and Basingstoke

Published in the U.S.A. by
Halsted Press, a division of
John Wiley & Sons, Inc., New York

Printed in Hong Kong

ISBN 0–470–20340–4

Library of Congress Cataloging-in-Publication Data
James, D. A.
    Radar homing guidance for tactical missiles.
    Includes bibliographies and index.
    1. Guided missles—Guidance systems.   2. Radar.
I. Title.
UG1310.J35   1986   623.4'519   86–9874
ISBN 0-470-20340-4

Contents

Preface

This book has its origin in a package of lectures given over a span of many years to the students of the Guided Weapons M.Sc. Course at the Royal Military College of Science, Shrivenham; the lectures formed part of a course of study covering the whole of guided weapons technology. Radar homing missiles have been in service in the armed forces of various countries throughout the world for at least 30 years and have been employed in many of the armed conflicts that have taken place since the end of the Second World War. Particularly noteworthy are the use of Surface to Air missiles in the Arab-Israeli conflicts and in Vietnam, and the devastating use of the Exocet sea-skimming anti-ship missile in the recent war in the Falkland Islands.

It may be asked why radar homing guidance has been singled out for special treatment; there are other forms of radar guidance and other forms of homing. The reason is that radar homing guidance combines two unique advantages: that of homing, in which the missile is theoretically certain to guide itself towards impact with the target, and that of radar which has the ability to penetrate rain, smoke, and dust. These advantages have encouraged much effort in development and in the growth of an extensive branch of radar technology which merits a treatment of its own.

Information pertaining to this subject has hitherto been rather scattered — in articles appearing in electronic engineering periodicals, in other works as part of a more general treatment, and in unpublished memoranda of a non-confidential nature within defence organisations. The purpose of this book is to collate this information sufficiently for the benefit of engineers working in this field; it is hoped that it will also prove to be of interest to engineers concerned with other branches of radar and telecommunications.

Treatment of the subject assumes no more than a knowledge of basic radar principles. It is mainly descriptive, supported by simple mathematical analyses and practical numerical examples; so far as security permits, the descriptions relate to actual systems. The introductory chapter gives a broad outline of the subject, including a general description of the homing process, operational aspects, and a selective catalogue of radar homers. It includes also a section on the basic principles of tracking radar upon which radar homing depends. Chapters 2, 3 and 4 describe features which are common to all applications; chapters 5, 6 and 7 deal with homing against airborne targets, and chapter 8 describes homing against slow-moving surface targets such as ships and vehicles, and includes a section on the specialized topic of homing on to enemy radar. Electronic counter-measures, which have already proved themselves to be a significant factor in the radar war, are described in the context of the particular applications mentioned. The book includes a glossary of those terms which are, on the whole, peculiar to this subject.

In conclusion, my thanks go to Professor C. J. Harris, Head of the School of Electrical and Electronic Science, Royal Military College of Science/Cranfield, for his encouragement, advice and provision of many useful facilities; to my colleagues for the provision of material and for constructive criticism; to Karen for the typing and to Les Judd for the many diagrams. Last, but not least, I wish to thank my former students for their interest and enthusiasm, which prompted me to explore the subject in depth.

The Ministry of Defence has given approval for this book to be published.

Glossary of Terms

All-the-way	Use of the same system of guidance throughout missile flight.
Autonomous	Missile can guide itself without further assistance from parent station.
Autopilot	See *Control system.*
Bistatic radar	Transmitting and receiving antenna are at separate locations.
Boost (phase)	Initial acceleration of missile to cruising velocity; short, a few seconds at most; severe, many 10s of *g*.
Chaff	Metallized strips of paper etc., resonant at radar wavelength; intended to confuse.
Conical scan	Method of tracking in angle, using a single rotating antenna primary feed.
Control system	Controls the flight and attitude of the missile, usually by means of appropriate aerodynamic surfaces.
Direct clutter	Breakthrough from transmitter (active) or rear reference (semi-active) into signal receiver; has zero Doppler shift in coherent system.
Ground spike	Doppler shifted signal due to ground echoes in the seeker antenna main beam.
Home-on-jam	Switching from active or semi-active to passive homing on a source of electronic counter-measure (ECM).
Homing eye	The seeker antenna system.
Homing head	The seeker.
Illuminator	Transmitter providing the radiation scattered from the target as an echo; can be in the seeker (active) or at the parent station (semi-active).

Incidence lag	Non-alignment of the missile longitudinal axis with the heading.
Inverse receiver	The Doppler shift on the target echo is transferred to the reference channel.
LATAX	Lateral acceleration (abbreviation).
Lock-on	Commencement of steady tracking of the target in all coordinates.
Mid-course guidance (phase)	A different and probably less precise form of guidance used during most of the missile flight.
Monopulse receiver	Radar receiver which derives a complete set of target co-ordinates from each pulse echo — that is, measurements are simultaneous.
Monostatic radar	Transmitter and receiver are co-located.
Multi-path	Propagation of radio waves by a direct path and one or more surface-reflected paths.
Phased array	Antenna in which the beam is steered electronically, without mechanical movement of the antenna.
Pitch	Movement of the missile or antenna in the local vertical plane (comparable to 'elevation').
Range gate	A gating pulse timed to coincide with the arrival of a pulse echo.
Rear reference	Direct transmission from the illuminator in semi-active homing to the missile in flight to provide synchronization.
Rear reference spillover	See *Direct clutter*.
Resolution cell	'Space' defined by the combined resolution in all co-ordinates.
Roll	Rotation of the missile about its longitudinal axis.
Seeker	The radar homing guidance installation in a missile.
Sequential lobing	Sequential sampling of the signals of a monopulse angle tracking antenna; equivalent electrically to conical scan.
Squint	Angle between sight line and missile longitudinal axis.
Terminal guidance (phase)	Precise guidance system taking over in the final stage of missile flight.
Threshold detector	Detector which takes cognizance of a signal only if its

amplitude attains a given level for a given minimum duration.

Velocity gate   Narrow band filter tuned to accept the Doppler frequency of the target echo.

Yaw   Movement in the local horizontal plane (comparable to 'azimuth').

List of Symbols

a	aberration
A	antenna aperture area
B	bandwidth
c	velocity of electro-magnetic waves in free space, 3×10^8 m s^{-1}
C	clutter power
d, d'	diameter; spacing
E	emissivity
f	frequency; function
F	Fourier transform
g	acceleration of gravity, 9.81 m s^{-2}
G	antenna gain
h	Planck's constant, 6.626×10^{-34}; height
J	received jamming power
k	an integer; constant of proportionality; Boltzmann's constant, 6.626×10^{-34} J K^{-1}
K	constant of proportionality
L	power loss factor
M	Mach number
n	an integer
N	noise power; an integer
NF	noise figure
P	transmitted power
r	radius; an integer
R	range
s	Laplace transform parameter
t	time
T	time constant; fixed time interval; temperature
T_0	ambient temperature, 290 K
u	velocity
W	fast Fourier transform (FFT) parameter
x, y, z	Cartesian coordinates
Y	transfer function

Greek

α	geometrical angle; radome aberration slope
β	geometrical angle

γ	geometrical angle
δ	geometrical angle; incremental
$\tan \delta$	loss tangent
Δ	difference (finite)
ϵ	radome aberration angle
ϵ_r	relative permittivity
ζ	damping parameter
η	efficiency
θ	geometrical angle; θ_3 antenna half-power beamwidth
λ	wavelength; proportional navigation constant
π	3.1418
σ	radar echoing area; standard deviation
σ_0	radar echoing area per unit area of clutter
σ'	radar echoing area per unit area of diffusing surface
Σ	sum
τ	pulse duration
ϕ	phase angle; geometrical angle
ψ	angle with respect to space axes; geometrical angle
ω	angular velocity; pulsatance $(2\pi f)$

Suffixes

a	atmospheric; aberrated sight line
c	carrier; rear reference spillover
C	clutter
D	boresight axis; Doppler
e	elapsed; earth
f	missile heading
g	glint
i	integrated
I	interferometer; illuminator
J	jammer
m	missile
n	natural
o	local oscillator
p	pitch
r	recurrence; receiver
R	receiver
s	target sight line; system
t	transmitted; target
T	target
y	yaw

List of Abbreviations

AAM	air to air missile
AC	alternating current
AFC	automatic frequency control
AFV	armoured fighting vehicle
AGC	automatic gain control
ARM	anti-radar missile
ASM	air to surface missile
CS	compressive strength
CW	continuous wave
DC	direct current
DFT	discrete Fourier transform
ECCM	electronic counter-counter-measures
ECM	electronic counter-measures
EM	electro-magnetic
ERP	effective radiated power
ESM	electronic support measures
FFT	fast Fourier transform
FM	frequency modulation
FMCW	frequency modulated continuous wave
FMICW	frequency modulated interrupted continuous wave
FS	flexural strength
GBJ	ground-based jammer
IF	intermediate frequency
IFA	intermediate frequency amplifier
LO	local oscillator
LOS	line of sight
MCOPS	millions of complex operations per second
MIC	microwave integrated circuit
PA	probability of acquisition
PN	proportional navigation
PRF	pulse recurrence frequency
PSR	phase-sensitive rectifier
RF	radio frequency
SAM	surface to air missile

SAW	surface acoustic wave
S:N	signal to noise ratio (power)
SOJ	stand-off jammer
SSJ	self-screening jammer
SSM	surface to surface missile
TGSM	terminally guided sub-munition
TIR	tracking and illuminating radar
TVM	target via missile
UTS	ultimate tensile strength
VCO	voltage-controlled oscillator
YIG	yttrium–iron–garnet

1

Basic Principles

1.1 Introduction

The frequent minor wars of the last two decades have brought to the fore the guided missile as a weapon against all types of military target. Notable are the successful use of surface to air missiles in the Israeli–Arab war of 1973, the sinking of the destroyer *Eilat* by a Russian-made anti-ship missile, and the extensive use of surface to air and air to air missiles in the Vietnam conflict; more recent are the extensive and devastating use of surface to air, air to air, and anti-ship missiles in the Falkland Islands war of 1982 and the successful use of surface to air missiles in the hands of small irregular forces.

In a free-flight weapons system, such as an artillery shell fired from a gun, the location and any movement of the target are measured as accurately as possible up to the moment of firing and a point of impact is calculated, based upon prediction of the movement of the target during the time of flight of the projectile. Two sources of major error exist in this process: firstly the target may change course or speed during the time of flight, and secondly the projectile may not follow the expected trajectory because of errors in the velocity and direction of projection and of unpredictable variations in the meteorological conditions prevailing over the flight path. A guided weapon system avoids both these sources of error by monitoring continuously the flight of the missile with respect to the position and movement of the target, thereby ensuring a much higher probability of a kill. The probability of destroying an airborne target by sustained gunfire is about 10 per cent at best; with a guided missile it can be as high as 50 per cent with a single shot.

The basic features of a system of guidance are shown in figure 1.1: a guidance computer computes the appropriate corrections to the missile trajectory, in

1

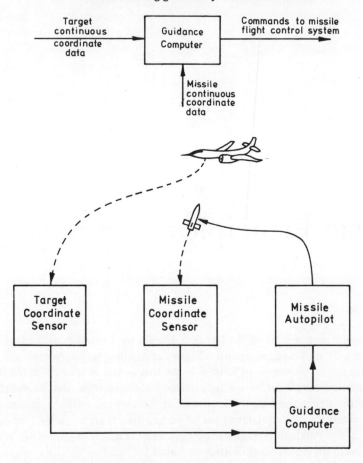

Figure 1.1   *Guidance principles*

response to the behaviour of the missile and target as recorded by sensors; the corrections are applied to the missile flight control system, or autopilot, which alters the flight path accordingly. For a target which is capable of movement, such as an aircraft, ship, or vehicle, the choice of sensor lies between radar, infra-red, and optical. Infra-red and optical sensors are capable of greater precision of angular measurement than radar, because of their much shorter wavelength, but radar has the advantage of much better penetration of mist, rain, dust, and smoke. Hence radar is favoured for all except short-range (a few km) guided weapons systems. There are several methods of using radar to guide a missile.[1] In command guidance, a radar based on the parent platform (ground, ship, aircraft, or vehicle) tracks the missile; a second, similar, radar tracks the target; a guidance computer compares the two sets of tracking data and issues the appropriate corrections via a radio link to the missile in flight. In beam riding guidance, a single parent radar tracks the target; the missile navigates itself up the radar beam by sensing its location with respect to the beam axis. In homing guidance, the missile carries its own tracking radar, known as a *seeker*, with its antenna in the nose. The seeker tracks the target and issues corrections to the autopilot, causing the missile to head continuously towards impact with the target.

The advantages of homing compared with the other methods are

(1) The precision of tracking improves as the missile closes upon the target and the range shortens, thus rendering the probability of success largely independent of the initial range of the target from the parent installation of the missile.
(2) The missile is potentially autonomous — that is, once launched it guides itself towards the target without further assistance from the parent installation, leaving the latter free to engage another target.

The disadvantages are

(1) An expensive guidance package is lost with each missile fired (although recent improvements in miniature circuit technology have reduced the expense of the guidance package).
(2) The guidance package occupies the prime location within the missile, displacing the warhead.
(3) The missile nose has to be a vulnerable and expensive radome (although better radome materials and smaller missile diameters have led to stronger radomes).

The advantages of homing are so important that it is now almost invariably chosen as the method of guidance for all but close support guided weapons — that is, weapons with a range of only a few kilometres.

Layout of a radar homing missile

Figure 1.2 depicts a typical layout; the radome, which is transparent to radio waves, protects the seeker antenna and forms the aerodynamic nose of the missile. The guidance system electronic circuits lie in the bay immediately behind. The rearward-facing antenna in the tail is for the passage of miscellaneous commands and information between the parent installation and the missile. Figure 1.3 is a photograph of a typical seeker antenna and its associated radome.

1.2 The homing process

The control of the missile trajectory is usually in two cartesian planes — horizontal or yaw, and vertical or pitch. The two planes are separate, hence there are two separate but similar guidance and control systems. There may also be a roll

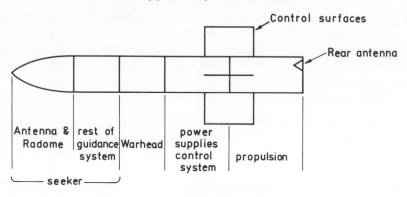

Figure 1.2   *Layout of a radar homing missile*

Figure 1.3   *Seeker antenna and radome (courtesy of Royal Military College of Science)*

control system to maintain the orientation in space of the transverse axes of the missile.

Figure 1.4 depicts the geometry of homing, in one plane only, for simplicity. The seeker antenna defines the sight line to the target, MT, by tracking in angular coordinates. From the tracking data, the missile is constrained to follow a 'proportional navigation' (PN) course, defined by the equation

$$\dot{\psi}_f = \lambda \dot{\psi}_s \qquad (1.1)$$

where $\dot{\psi}_f$ is the rate of change of missile heading with respect to inertial axes, $\dot{\psi}_s$ is the rate of change of sight line, also with respect to space axes, and λ is a constant of proportionality known as the *navigation constant*. The lower limit,

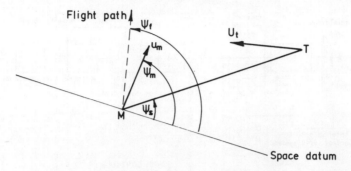

Figure 1.4 *Proportional navigation*

$\lambda = 1$, leads to a tail chase of a moving target, demanding an infinite rate of turn of missile heading — that is, infinite lateral acceleration (LATAX), at the end of the engagement (see figure 1.5). The upper limit, $\lambda = \infty$ represents the constant bearing course in which the missile flies in a straight line to a predicted point of impact; this is an ideal course but is unrealizable in practice. The practical value of λ is just high enough to eliminate the effects of misaiming early in the flight so that, barring late manoeuvres by the target, the missile should be tending to follow a constant bearing course as it approaches impact. A low value of λ, typically slightly greater than 2, is suitable for engaging slow-moving or stationary targets; a high value, 5 or more, is required for engaging fast-moving manoeuvrable airborne targets.

Figure 1.6 shows a block diagram of the complete homing guidance and control process in one plane. The seeker antenna tracks the target; because the seeker antenna servo system is error-actuated, there is an error voltage unless the target is stationary with respect to the missile. In conventional tracking radar, this error voltage drives the position servo motor directly. In homing, the pointing error voltage is compared with the output of the rate gyroscope attached to the seeker gimbal and the difference is used to drive the servo motor. This ensures that the antenna rate is measured with respect to an inertial axis, even though the servo motor drives the antenna with respect to the missile. The point-

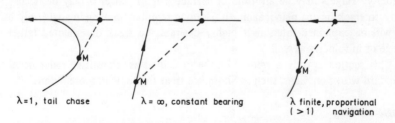

Figure 1.5 *Navigational courses*

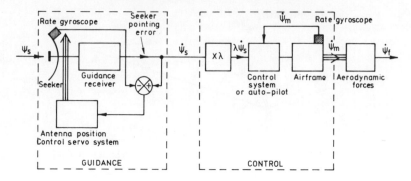

Figure 1.6   *Guidance and control process*

ing error voltage is used as a measure of $\dot{\psi}_s$. Thereafter $\dot{\psi}_s$ is multiplied by λ and passed to the missile control system as a demand for a rate of turn. This rate of turn $\dot{\psi}_m$ is monitored by another rate gyroscope, thus closing the control loop. Strictly $\dot{\psi}_m$ is not the same as the rate of change of missile heading, $\dot{\psi}_f$, because of aerodynamic incidence lag; since this distinction is not usually of direct importance to the guidance system, it will be ignored henceforth. In a two-plane homer, the guidance and control system is duplicated in the second plane; in a single-plane homer, such as a sea-skimming missile, homing is in the yaw plane only — in the vertical plane the missile is constrained to fly at a pre-set height above the sea by means of an altimeter associated with the pitch control system. This account is merely an outline of the overall guidance and control process; further information about control systems and associated aerodynamics can be found in relevant textbooks, such as Garnell[2].

1.3 A note on tracking radar

The purpose of conventional tracking radar is to provide continuous precise measurement of the coordinates of a selected target; these coordinates are usually azimuth, elevation, and range, though sometimes they include relative velocity. Targets may be airborne or surface; for the latter it may be necessary only to track in one plane, azimuth. Homing requires tracking in angle only but, as will be explained later, it is highly desirable to track the selected target in range or in velocity as well.

This section is only a *résumé* of the principles as relevant to radar homing; standard works on radar, such as Skolnik[3], treat the topic in more detail.

Angle tracking

Precise angle tracking depends upon the use of a narrow moving antenna beam; for single-plane tracking, the beam need be narrow in that plane only; tracking in both planes requires a pencil-shaped beam, which is often generated by an antenna with a reflector in the form of a paraboloid of revolution, with primary source at the focus. The width (that is, between half-power points), θ_3, of the beam in radians is given by the formula $k\lambda/d$, where λ is the wavelength, d is the antenna diameter in the same units, and k is a constant lying between 1.2 and 1.5. For example, to generate a beam 17 mrad ($1°$) wide at a wavelength of 30 millimetres requires a reflector diameter of between 2.1 and 2.6 metres. For tracking purposes, the antenna must generate a voltage that is proportional in both magnitude and sign to the misalignment of the antenna axis from the sight line to the target; in order to do this, there are two beams, L and R, in the tracking plane, say azimuth, instead of just one, whose axes are displaced equally and on opposite sides of the antenna axis (figure 1.7a). The angle of displacement, or squint, is about $\theta_3/2$. For two-plane tracking there is a similar pair of beams, U and D, in the elevation plane. Consider tracking in azimuth: the echo signal voltage from the target T, lying to the left of the antenna bore-sight axis OZ, is stronger in beam L than in beam R, and the magnitude and sign of the difference Δ between the two signals gives a direct measure of the pointing error θ_T (figure 1.7b). After processing in the receiver and conversion to DC, this difference voltage is used to energize the antenna azimuth position control servo motor, thus driving the antenna to reduce the pointing error to zero.

Both pairs of beams can be generated simultaneously by means of a multiple primary antenna source associated with a combining network, or hybrid; the name of this arrangement is *static split* or *monopulse*, because each echo yields complete angle-tracking data. Monopulse provides good tracking but is complicated; a simpler method is to generate a single offset beam which rotates around the antenna axis, sampling the echo signal strength at each of the four cardinal points (figure 1.8). The scanning beam traces out an imaginary cone, hence the name *conical scan*. The rotating offset beam is created by spinning the primary source, which is itself offset, about the focus; in single-plane tracking, the primary source oscillates in a straight line. The spin rate must be high, so that random changes of signal strength do not affect the validity of the comparison — several hundred hertz is typical. Nevertheless, the sequential nature of the

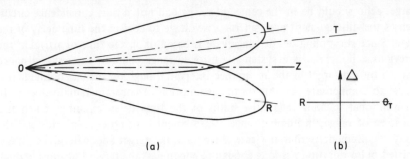

(a)   (b)

Figure 1.7   *Two-beam angle tracking*

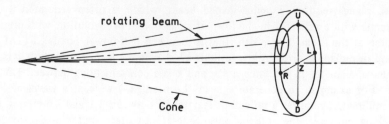

Figure 1.8   *Conical scan*

measurements renders the method vulnerable to interference and to simple ECM, hence monopulse is almost always preferred to conical scan in homing seekers.

Tracking in range

In conventional tracking radar, the primary purpose of tracking in range is to provide continuous measurement of range; in homing guidance, the primary purpose is to segregate the selected target echo from the echoes from all other targets lying in the radar beam, in particular from clutter. The techniques are, however, the same.

Most range-tracking radars transmit a continuous series of short pulses; assuming that the receiver and transmitter are at the same location (monostatic radar), the echo of each pulse returns from a target at a range R after an elapsed time $T = 2R/c$ (figure 1.9) where c is the velocity of EM waves. T is short, of the order of tens or hundreds of microseconds; a useful conversion is 1 microsecond $\equiv$ 150 metres. The pulse repetition frequency (PRF) is sufficiently high to provide adequate continuity of measurement, but it is also sufficiently low to allow echoes to return from targets at the extreme operational range before the next pulse is transmitted; typical values are several hundred or several thousand hertz. In order to track in range, a time delay circuit (figure 1.10) generates a range gate pulse of duration about twice that of the transmitted pulse, timed to coincide with the arrival of the next echo from the selected target (figure 1.9). The target echo should lie in the centre of the gate; if not, a time coincidence circuit senses the misalignment and generates a voltage to correct the time delay of the gate. Since the timing of the range gate is based upon the time of arrival of the previous echo, there is a systematic error in range equal to the radial distance moved by the target in the recurrence interval; though the error is not large it is usual to compensate for this on the assumption of constant radial velocity. In conventional tracking radar, the status of the time delay circuit is taken as a measure of range; in homing range measurement is of no consequence and the only requirement is that the gate shall track the target adequately. The transmitted pulse duration τ is long enough to encompass the target but short enough to provide adequate resolution from unwanted targets; a typical value is 0.5 μs

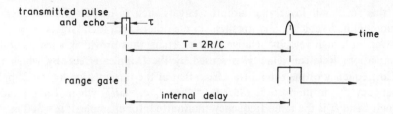

Figure 1.9   *Range gating waveforms*

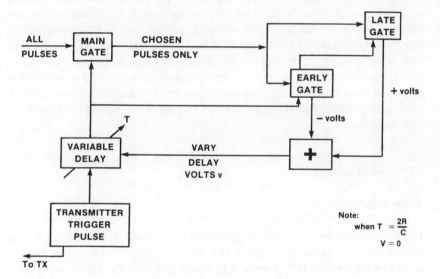

Figure 1.10   *Time delay circuit*

(75 m range resolution). Initially, the range gate is placed over the target echo by making use of range information from a separate target indicator such as a surveillance radar. If this range information is too coarse, the range gate has to search an ambit until it acquires the target. It is possible to make use of other forms of ranging modulation; the only requirement is that the spectral bandwidth B of the modulated transmission shall be at least equal to the reciprocal of the time resolution required; for example, any modulation bandwidth of 2 MHz will yield a range resolution of 75 m. Nevertheless, practical considerations usually restrict the choice of other forms to frequency and phase modulation.

Tracking in relative velocity

Range gating can usually provide adequate discrimination between large surface targets, such as ships and tanks, and surrounding terrain (clutter); it can rarely do this for small low-flying airborne targets such as modern fighter aircraft, remote-piloted vehicles, and missiles. On the other hand, such targets are usually moving with high velocity relative to the terrain, thus providing a means of discrimination. Relative velocity is sensed by the Doppler effect, by which the radio frequency of the echo differs from that of the transmission by the Doppler frequency f_D; in monostatic radar $f_D = 2u \times f_c/c$, $=2u/\lambda$, where u is the radial velocity and f_c is the radio frequency transmitted; for example, if $u = 300$ m s^{-1} and $f_c = 10$ GHz, then $f_D = 20$ kHz. The selected target is segregated by means of a narrow bandpass filter, or velocity gate, tuned to the target's Doppler frequency, and an automatic frequency control system maintains correct tuning in the face of changes in radial velocity of the target; as in range gating, the velocity gate is set initially from external data. A typical gate bandwidth is 500 Hz, giving a velocity resolution of 15 m s^{-1} at a radio frequency of 10 GHz. With velocity gating, it is sufficient to employ a CW transmission; however, in a very congested target environment, range gating may be required as well; in this case there must be a ranging modulation present. The most usual modulation is pulse, giving rise to pulse-Doppler, though frequency modulated CW (FMCW) is used sometimes.

1.4 Types of homing

Active homing

The missile contains the complete radar, with transmitter. Acting homing is fully autonomous, as shown in figure 1.11a, but it has the disadvantages that an expensive transmitter is lost with each round and that the maximum range is short, about 10 km, because of the small antenna and the limited transmitter power. Hence active homing is confined to short range 'all-the-way' homing systems, or to 'terminal' guidance in longer-range systems which use a different, less precise, form of 'mid-course' guidance.

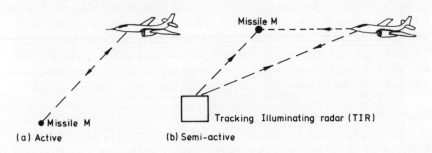

Figure 1.11   *Types of homing guidance*

Semi-active homing

It is possible to overcome these disadvantages by separating the transmitter (here known as the 'illuminator') from the missile and installing it permanently at, or near, the launch point, as shown in figure 1.11b. The effective radiated power (ERP) (product of radiated power x antenna gain) can be much greater, permitting much longer range. The price is the loss of autonomy — that is, the ability of the missile to function on its own once it has been launched. It is possible to restore a measure of autonomy by using a multi-target illuminator, such as a phased-array radar, but at the cost of complexity and financial expense. Early homers used the same radar for tracking and for illuminating the target; figure 1.12 is a photograph of such a radar for a medium-range (about 30 km) surface to air missile (SAM) system. However, the two functions are not entirely compatible and modern practice is to use a separate illumination beam, sharing the antenna with the tracking beam. Figure 1.13 is a photograph of the radar for a shipborne medium-range SAM which uses this principle. In addition to illumination of the target, the guidance receiver also requires direct information about the frequency of the illumination beam and of any modulation. Some systems rely on pre-launch settings but it is more usual to convey the information continuously to the missile in flight through a 'rear reference' link. The rear reference transmission is usually from a separate antenna, which follows the main illumination antenna, to an antenna in the rear of the missile. The rear reference transmitting antenna has a moderate gain and a beam which is wide enough to encompass manoeuvres of the missile in flight; in figure 1.12 the rear reference

Figure 1.12   *Tracking and illuminating radar (courtesy of Royal Military College of Science)*

Figure 1.13 *Tracking and illuminating radar, single antenna (courtesy of Royal Military College of Science)*

antenna is the flat antenna alongside the paraboloid. The rear reference antenna in the missile is quite simple, say a low-gain horn, and has a wide field of view. Where it is difficult to accommodate a separate rear reference transmitting antenna, as in an aircraft, it may be sufficient for the missile to rely on the edge of the main beam or on a side-lobe.

A mention was made above of supplying a measure of autonomy by the use of a multiple target tracking and illuminating radar; figure 1.14 shows the US Patriot system radar, which can engage several targets simultaneously through use of a phased-array. (The Patriot system uses, in fact, a variation of semi-active homing known as 'target via missile', TVM, which is described in section 6.7.)

Passive homing

If the target itself is radiating at a suitable frequency, the missile can home on to the source without the need for illuminating the target. For example, if the target attempts to jam the illumination of an active or semi-active homer, the guidance receiver can switch to 'home on jam'. The specialized anti-radar (ARM) missile homes on to the source of an enemy radar transmission, destroying the antenna. It might be possible to home on to radiocommunications transmissions but the frequencies employed are usually too low for accurate angle tracking. However, RAM, an anti-ship missile being developed in the USA, uses homing on to general radio-frequency emissions for mid-course guidance.

A much more radical approach is to make use of the part of the spectrum of

Figure 1.14 *Patriot radar (courtesy of the Raytheon Company)*

natural thermal radiation of the target that falls within the radio-frequency band. This is akin to infra-red homing but it has the advantage that the radiation penetrates fog, smoke, and rain better. Unfortunately, the spectral radiant emission at the relatively long radar wavelengths is very low; nevertheless, it is possible to home on to a large vehicle or on to an aircraft from a distance of 1 or 2 km by working at the shortest possible wavelengths, in the millimetre (mm) wave band, and by using a sensitive radiometric receiver. This may seem to be a modest achievement, but the advantages of being passive and of requiring no cooperation from the target make the concept worth consideration for terminal phase guidance.

1.5 Examples of radar homers

Most of the radar systems in existence are the products of France, the UK, the USA, and the USSR. Table 1.1 gives a selection, together with references in open literature. Systems are usually classified operationally according to the following list

Air to air	AAM
Air to surface	ASM
Surface to air	SAM
Surface to surface	SSM

A selection of photographs of missiles listed in table 1.1 appears in figures 1.15 to 1.20. *Jane's Weapon Systems* gives a comprehensive catalogue of missiles and the *International Defense Review* (IDR) is a valuable source of information about new developments.

Table 1.1. A selection of radar homing missiles

AAM (air to air missiles)

AA series	USSR	
Phoenix[4]	USA	Active and semi-active versions
Skyflash[5,6]	UK	Semi-active
AMRAAM	USA/NATO	Active

ASM (air to surface missiles)

Exocet[7,8]	France	Active terminal phase, anti-ship sea skimmer
Kennel (AS1)	USSR	Active terminal phase, anti-ship
Sea Eagle[7]	UK	Active terminal phase, anti-ship sea skimmer
Sea Skua[9]	UK	Semi-active, helicopter-launched, anti-ship sea skimmer
Wasp[10]	USA	Active, battlefield, anti-vehicle

SAM (surface to air missiles)

Bloodhound Mk II	UK	Semi-active, land-based, obsolescent
Goa (SAN 1)	USSR	Active terminal, shipborne
Hawk	USA	Semi-active, land-based
NATO 6S[4]	USA/NATO	Semi-active mid-course, fully active terminal, shipborne
Patriot[4]	USA	Semi-active (TVM), multi-target capability, land-based
SA6 (Gainful)[11,12]	USSR	Semi-active, land-based
Sea Dart	UK	Semi-active, shipborne
Quick Shot-Sure Shot[13]	USA	Active terminal (mm wave), anti-ballistic missile
Sea Sparrow[6,14]	USA/NATO	Semi-active, shipborne
Standard[4]	USA	Active terminal, shipborne

SSM (surface to surface missile)

Assault Breaker[15]	USA	Battlefield, anti-vehicle, active terminally guided sub-munitions (TGSM), also passive radiometric
Merlin	UK	Guided mortar bomb, mm wave
MRLS (multiple rocket launch system)		Battlefield, anti-vehicle, active, mm wave TGSM
Exocet	France	As ASM, also submarine launched
Harpoon[16,17]	USA	Active terminal, anti-ship sea skimmer, submarine launched

SSN series	USSR	Active terminal

ARM (anti-radar missiles)

Martel	France/UK	Air launched
Shrike	USA	Air launched
Harm[10]	USA	Air launched

Several USSR SSM and ASM have anti-radar versions

Figure 1.15   *Hawk missiles (courtesy of the Raytheon Company)*

Figure 1.16   *Skyflash on Tomada (courtesy of British Aerospace Dynamics Group)*

Figure 1.17   *(a) Sea Skua on Lynx helicopter; (b) Sea Eagle on Sea Harrier (both courtesy of British Aerospace Dynamics Group)*

Figure 1.18 *SA6 – Straight Flush radar and missiles (courtesy of Jane's Publishing Company)*

Figure 1.19 *Alarm ARM (courtesy of British Aerospace Dynamics Group)*

Figure 1.20   *Merlin anti-tank guided bomb (courtesy of British Aerospace Dynamics Group)*

1.6 Operational sequences

This section gives an outline of how two quite different systems engage their respective targets. The first is an all-the-way semi-active homing SAM system, depicted in figure 1.21. The sequence of engagement is as follows

(a) Surveillance radar detects the target, and passes coordinates to control centre.
(b) Control centre assesses threat, decides to engage, and passes coordinates to the Tracking and Illuminating Radar, TIR.
(c) TIR searches and acquires the target, and then tracks.
(d) TIR passes accurate tracking data to control centre.
(e) Control centre computes initial launch angles, target sight line and Doppler frequency, and passes data to seeker.
(f) When the target is in range of the seeker, the control centre launches the missile; during the short boost phase there is no guidance and control.
(g) At the end of boost, the seeker acquires the target and tracks.
(h) The missile comes under full guidance and control.

The second example is an anti-ship sea skimmer, surface launched, with inertial mid-course guidance and terminal active radar homing. The sequence of engagement is as follows (figure 1.22)

(a) Target is detected from over the horizon, say by sonar or by airborne surveillance.
(b) Target coordinates are set into the inertial guidance system, allowing for predicted target movement during the mid-course phase.
(c) Missile is launched and descends to a sea-skimming trajectory, a few metres above the sea, before it crosses the target's horizon; sea-skimming height is maintained by altimeter.
(d) When within radar distance, the seeker searches for the target in yaw and range.
(e) Seeker acquires the target and tracks; missile comes under full guidance and control.

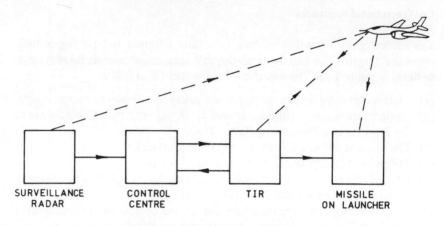

Figure 1.21   *Semi-active SAM system*

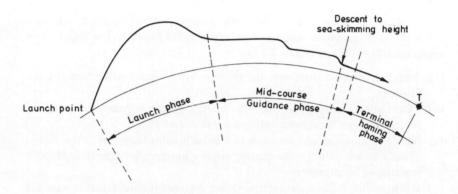

Figure 1.22   *Sea-skimmer missile trajectory*

References

1. Maurer, H. A., 'Trends in missile guidance', *International Defence Review*, 9/1980, p. 1415.
2. Garnell, P. G., *Guided Weapon Control Systems*, 2nd edn, Pergamon Press, 1980, Chapters 8 and 9.
3. Skolnik, M. I., *Introduction to Radar Systems*, 2nd edn, McGraw-Hill, 1980, Chapter 5.
4. Hewish, M., 'Tactical missile survey no. 2, air targets', *International Defence Review*, 9/1980, p. 1431.
5. 'The Skyflash air to air missile', *International Defence Review*, 6/1979, p. 1951.

6. 'The Sparrow family', *Defence*, March 1981.
7. Hewish, M., 'The Sea Eagle anti-ship missile', *International Defence Review*, 7/1984, p. 937.
8. 'Inside the Exocet', *Defence Electronics*, 8/1982, p. 46.
9. Preston, A., 'Sea Skua', *Defence*, Sept. 1980, p. 681.
10. *Flight International*, 8 Jan. 1983, pp. 91–3.
11. 'SA6, Arab ace in the 20-day war', *International Defence Review*, 6/1973, p. 781.
12. 'New details on the SA6', *International Defence Review*, 4/1974, p. 529.
13. *Aviation Week*, 24 Jan. 1983, pp. 30–31.
14. 'Sea Sparrow', *NATO's 15 Nations*, Dec. 1981 and Jan. 1982.
15. Hewish, M., 'Tactical missile survey no. 1, ground targets', *International Defence Review*, 6/1980, p. 851.
16. 'The Harpoon missile system', *International Defence Review*, 1/1975, p. 61.
17. Preston, A., 'The deadly Harpoon', *Defence*, May 1981, p. 307.

2

Types of Radar Transmission

2.1 Introduction

Various types of radar transmission have been mentioned in connection with tracking radar in section 1.3. The aim of this chapter is to explain the choice of transmission for homing guidance and to describe in outline the corresponding types of seeker.

Although the number of possible radar transmissions is large, practice tends to limit them to two main types – pulse and continuous wave (CW). Historically, the first is pulse, which yields range (or time delay) information by timing the arrival of the received pulse envelope from a target. In a homing missile, this information is used to provide a 'range gate' to segregate the wanted target from all other targets, clutter etc. in the antenna beam that are not at the same range as the target.

Continuous wave-Doppler yields closing velocity information but tells nothing about elapsed time or range. The Doppler frequency shift of the carrier of the received signal from a target is used to provide a 'velocity gate' to segregate the wanted target from others having a different closing velocity. In particular, this type provides good discrimination between a moving target and echoes from the ground or sea (clutter). Range, or elapsed time, can be measured by frequency modulating the CW (FMCW), but this is usually ancillary to the main function of providing a velocity gate.

Pulse-Doppler is also basically a Doppler system. However, pulsing of the carrier offers certain advantages over CW; in particular, it can separate the received signal in time from the transmitted or rear reference signal, so facilitating the sharing of a single antenna.

22

2.2 Pulse

This leads to a system which is basically similar to conventional pulse tracking radar. The unwanted target is selected and tracked automatically in range (or elapsed time) by an electronic range gate. Only the range-gated target is allowed to pass to the angle-tracking circuits which govern the pointing of the homing eye.

Active

This is identical to the monostatic pulse ranging radar described in chapter 1. Figure 2.1 shows: (a) the missile (M) — target (T) geometry, (b) the block diagram of the system, and (c) the relative timings of the transmitter, target echo, and range gate pulses. As is typical of automatic pulse ranging systems, the width of the range gate is about twice that of the target echo. The transmit/receive switch (T/R), or duplexer, permits sharing of the antenna on a basis of time.

Semi-active

The physical separation of the transmitter–illuminator, I, and the receiver in the missile, M, as shown in figure 2.2a, complicates the timing of the range gate within the missile. The range gate is initiated by an internal timing pulse which starts the timing circuit; this timing pulse, which must be synchronized to the PRF of the illuminator transmission, creates the timing sequence shown in figure 2.2b; R_1, R_2, and R_3 are the ranges shown in figure 2.2a.

PRF synchronization is usually provided through a rear reference transmission directly from the illuminator to the missile in flight; this transmission also provides a link between the carrier frequency of the illuminator transmission and the frequency generated by the microwave local oscillator for the superhet receiver within the missile; a conventional automatic frequency control circuit controls the local oscillator to maintain the correct IF for the missile receiver. The rear reference link can be omitted altogether provided that

(a) The illuminator PRF and the PRF generated within the missile are sufficiently stable with respect to each other. This usually means careful control of each by a crystal oscillator.
(b) The illuminator carrier frequency and the microwave local oscillator frequency are sufficiently stable with respect to each other to ensure that the target echo signal remains within the fairly wide pass band of the missile IF amplifier.

The earliest SAM and AAM systems used plain pulse transmission but they ran into the usual difficulty of target echo masking by ground clutter at the same range when trying to engage low-flying targets. Pulse is now confined for use against surface targets such as ships and vehicles, which are too slow to give

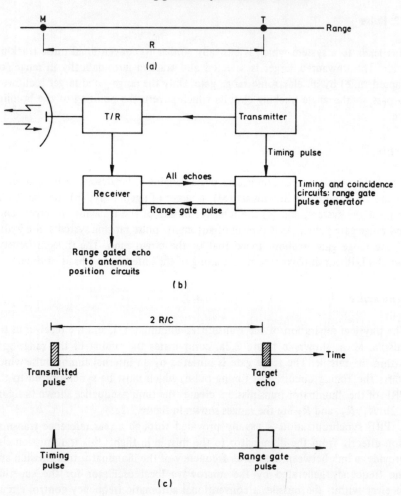

Figure 2.1   *Fully active pulse radar homing guidance system – range gating*

an appreciable Doppler frequency shift but which are large enough to stand out against clutter.

2.3 CW-Doppler

It has already been mentioned that pulse systems are vulnerable to ground, sea, or even weather clutter at the same range as the target. Doppler systems take advantage of the fact that airborne targets near the ground or sea are nearly always moving fast relative to it. This relative motion can be detected by the

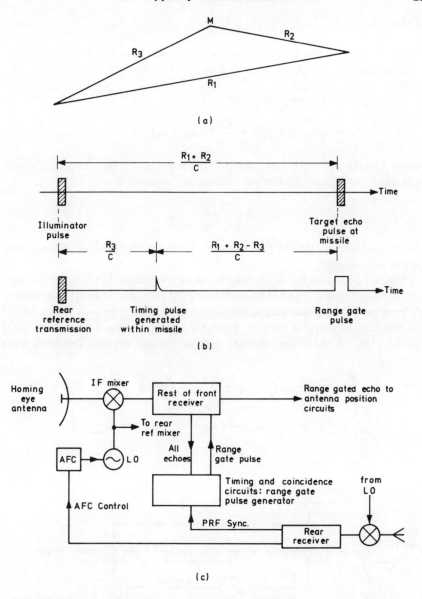

Figure 2.2   *Semi-active pulse radar homing guidance system – range gating*

Doppler frequency shift of the carrier of the target echo relative to that of the clutter; this Doppler shift can then be used to distinguish the wanted target from the unwanted clutter.

The basic Doppler system uses CW transmission since that is sufficient to extract the Doppler shift.

Active

Figure 2.3a shows the geometry of a simple situation in which the missile, M, and the target, T, are flying towards each other at velocities u_m and u_t respectively.

The target Doppler frequency shift is

$$f_{\mathrm{DT}} = \frac{+2f_c}{c} \, (u_m + u_t)$$

where f_c is the transmitted frequency. If the target is near the ground the missile will also receive ground clutter but its Doppler frequency is

$$f_{\mathrm{DC}} = \frac{+2f_c}{c} \, u_m$$

The difference between the two is usually adequate for rejection of the clutter by the velocity gate.

Figure 2.3b shows the block diagram of such a system. The Doppler frequencies of all echoes are obtained by subtracting the constant transmitted frequency f_c from the carrier frequencies of the echoes; the wanted target echo is segregated on the basis of Doppler shift by a narrow band pass filter (the velocity gate) before going to the seeker antenna position control circuits. The filter must

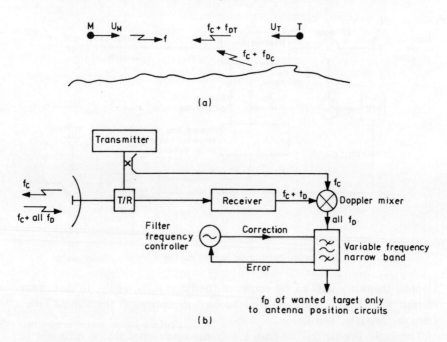

Figure 2.3   *Fully active CW-Doppler homing guidance system – velocity gating*

follow changes in Doppler frequency of the wanted target, just as the range gate in a pulse system follows changes in time of arrival or range. The mid-frequency of the filter is variable, controlled by a frequency controller. The velocity gate circuits sense any difference between the Doppler frequency of the gated target echo and the mid-frequency of the filter, using this difference to cause the frequency controller to retune the filter to eliminate error.

Note that it is possible to employ a single antenna by using a passive directional coupler as duplexer (transmit/receive). However the isolation is low, limiting the safe transmitter power, unless an elaborate breakthrough cancelling system is employed.

Semi-active

As in pulse, the situation is complicated by the separation of the transmitter-illuminator and the missile. Figure 2.4a shows the geometry for the simple case of the illuminator, I, the missile, M, and the target, T, all in a straight line with M and T flying towards each other with velocities u_t and u_m respectively.

If the transmitted frequency is f_c

Frequency re-radiated by T $\qquad\qquad = f_c \left(1 + \dfrac{2u_t}{c}\right)$ (2.1)

Frequency picked up by homing eye of M $\qquad = f_c \left[1 + \left(\dfrac{2u_t}{c} + \dfrac{u_m}{c}\right)\right]$

(2.2)

Frequency picked up by rear reference antenna of M $= f_c \left(1 - \dfrac{u_m}{c}\right)$ (2.3)

The guidance receiver takes the difference between (2.2) and (2.3), which is then known to the seeker as the *Doppler frequency* f_D of the target

$$f_D = \frac{2f_c}{c}\,(u_t + u_m)$$ (2.4)

If the target is flying low over the ground, the seeker will see clutter of Doppler frequency given by

$$f_{DC} = \frac{2f_c}{c}\,u_m$$ (2.5)

As in an active system, the difference between the two is usually sufficient for rejection of the clutter.

Because of the narrow effective bandwidth of a Doppler guidance receiver, it has until recently been considered essential to provide a rear reference in order to extract the Doppler frequency. Improvements in the frequency stability of the illuminator and of the local oscillator now make it possible for the rear reference to be dispensed with.

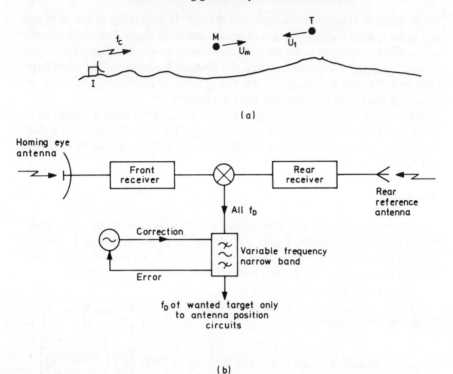

(a)

(b)

Figure 2.4   *Semi-active CW-Doppler homing guidance system — velocity gating*

Pulse-Doppler

The methods of extracting and tracking the Doppler frequency are the same as in CW-Doppler. Pulse-Doppler also provides the opportunity of gating the target in another coordinate — range; this is of particular value when the target density is high. Another advantage is that it is possible for the transmitter and the receiver of an active homer to share a single antenna on a time basis. The disadvantage of pulse-Doppler is the presence of ambiguities in range and in velocity indication.

Frequency modulated CW (FMCW)

Theoretically, this can provide velocity and range gating to the same standard as can pulse-Doppler. Unfortunately, practical difficulties tend to arise through the use of the same analogue — frequency — for both coordinates; furthermore, duplexing difficulties remain. So far FMCW has been used only as a fairly crude adjunct to CW for aiding the rejection of clutter when target and clutter Doppler frequencies are merging, as at a crossing-point. To minimize the risk of confusing the ranging and Doppler frequencies, the modulation frequency is chosen to be considerably less than the lowest Doppler frequency.

Frequency modulated intermittent CW (FMICW)

This seeks to provide both range and velocity discrimination without ambiguities, yet to retain the time-sharing provided by a pulse waveform. The CW is chopped at a rate sufficiently high to avoid velocity ambiguities, while the FM is at a sufficiently low rate to avoid range ambiguities. Good frequency stability is essential to avoid cross-coupling between velocity and range measurements, or gating, and careful choice of modulation frequency is necessary to avoid the un-expected ambiguities which tend to arise with elaborate modulations.

3

Range Performance

3.1 Introduction

An important characteristic of a missile system is the maximum range at which
the guidance receiver can acquire a given type of target. Since complete certainty
can never be guaranteed, an acceptable minimum probability of acquisition or
detection is also specified. Range performance depends upon a number of
external factors, as well as internal ones, such as meteorological conditions; these
factors must also be specified. The radar range equation is the usual vehicle for
evaluating range performance. A full discussion and detailed derivations can be
found in standard works such as Skolnik[1] but a simple derivation follows for the
benefit of those readers to whom the concept may be unfamiliar.

Assume that the transmitter and receiver are colocated, that is the radar is
monostatic. If the transmitted power is P_T and the boresight axis of the transmitting antenna is pointing directly at the target, the power density incident on
the target is

$$P_T \, G_T / 4\pi R^2$$

where G_T is the antenna gain and R is the range to the target. Most of the radiation intercepted by the target is scattered, some back towards the radar. The
measure of the proportion of the radiation scattered back to the radar is the
radar cross-section or the equivalent echoing area of the target, given the symbol
σ. It is defined as the effective power per unit solid angle scattered back to the
receiving antenna divided by the power density incident on the target. σ is a
property of the target size, shape and aspect, of the target material and of the
plane of polarization of the incident radiation with respect to the attitude of

the target. From this definition it follows that the power density incident on the receiving antenna is

$$\frac{P_T \, G_T}{4\pi R^2} \times \frac{\sigma}{4\pi R^2}$$

Assuming that the boresight axis of the receiving antenna is also pointing at the target, the power S delivered to the receiver input is given by a simple form of the range equation

$$S = \frac{P_T \, G_T \, \sigma A_R}{(4\pi)^2 \, R^4} \qquad (3.1)$$

where A_R is the effective aperture area of the antenna.

In most pulse radars the transmitter and receiver share a common antenna so that

$$G_T/4\pi = A_R/\lambda^2$$

giving two more alternative forms

$$S = \frac{P_T \, (G_T)^2 \, \lambda^2 \sigma}{(4\pi)^3 \, R^4} \qquad (3.2a)$$

$$S = \frac{P_T \, A_R^2 \, \sigma}{4\pi \, \lambda^2 \, R^4} \qquad (3.2b)$$

The first form is useful where the beam shape, and hence the antenna gain, is pre-determined; the second where the antenna size is restricted.

The received signal to noise ratio, $S{:}N$, is usually of more interest than just S. The effective value of N is the thermal noise power in the noise bandwidth (B_n) of the receiver multiplied by the receiver noise figure (NF) so that

$$N = (NF)kT_0 B_n$$

where k is Boltzmann's constant and T_0 is the ambient temperature (in K). A convenient formula is that when T_0 is 290K, $kT_0 = 4 \times 10^{-21}$ W Hz^{-1}. Usually B_n differs only slightly from the signal bandwidth B and the latter is usually adopted. Substituting the expression for N in equation 3.1 gives

$$S{:}N = \frac{P_T \, G_T \, A_R \, \sigma}{(4\pi)^2 R^4 \, (NF)kT_0 B} \qquad (3.3)$$

Often the maximum range R_{max} at which a target can be detected or acquired is the quantity required; this is given by

$$R_{max}^4 = \frac{P_T \, G_T \, A_R \, \sigma}{(4\pi)^2 \, (NF)kT_0 B \, (S{:}N_{min})} \qquad (3.4)$$

where $S{:}N_{min}$ is the minimum detectable at the receiver output. Since noise is a statistical quantity, $S{:}N_{min}$ is not a fixed quantity but depends, among other things, on the required probability of detection of the target.

It is also necessary to allow for miscellaneous losses such as atmospheric attenuation; these are incorporated by inserting a loss factor L in the denominator of the equation.

If the radar is bistatic, the range transmitter–target, R_1, is not equal in general to the range target–receiver, R_2, hence equation 3.1 is rewritten

$$S = \frac{P_T \, G_T \, A_R \, \sigma}{(4\pi)^2 R_1^2 R_2^2} \tag{3.5}$$

σ is also not necessarily the same as for the monostatic configuration; some information is available about bistatic values of σ and this should be used where possible.

The equations derived apply equally to pulse and to CW, provided that quantities are defined appropriately. In pulse radar, if P_T is the pulse power then B must be that of the IF amplifier. In this case B is wide, typically 1/pulse duration. In CW radar B is the width of the Doppler filter. In pulse-Doppler radar it is best to take P_T as the average power and B, again, as the Doppler filter bandwidth.

The purpose of this chapter is to relate the radar range equation to homing, giving examples of performance likely with current seeker capabilities.

3.2 Active homing

This is an example of monostatic radar for which the equation for the maximum acquisition range, R_{max}, is, from equation 3.4

$$R_{max}^4 = \frac{P_I \, \sigma A_m^2}{4\pi\lambda^2 L \, S_{min}} \tag{3.6}$$

where $S_{min} = NF \, (kT_0B) \, (S{:}N)_{min}$

P_I is the transmitted or illuminator power and A_m is the effective aperture of the antenna; other quantities are as in equation 3.4.

This form of the equation contains the aperture area as the antenna parameter because the antenna diameter is governed by the missile diameter. As an example, assume the following characteristics for a CW seeker for engaging airborne targets

P_I		50 W
σ		1 m^2
antenna diameter, d	0. 3 m	
aperture efficiency, η	0.35	
$A_m = \eta\pi d^2/4$		0.025 m^2
λ		22 mm

L	4 dB
NF	8 dB
B	1 kHz
$(S/N)_{min}$	+6 dB
giving R_{max}	12 km

This is a typical value, confirming that the performance of an active seeker is essentially short range. The calculation would be similar for pulse, but would be based upon pulse power and the much wider bandwidth associated with pulse transmission. Maximum range attainable is similar.

3.3 Semi-active homing

This is an example of bistatic radar, in which transmitter and receiver are separated. The relevant range equation is, from equation 3.5

$$(R_1^2 R_2^2)_{max} = \frac{P_I \, \sigma G_I \, A_m}{(4\pi)^2 \, L \, S_{min}} \qquad (3.7)$$

where R_1 = range illuminator–target and R_2 = range target–missile. If differs from equation 3.6 in that the relevant transmitting antenna parameter is G_I, because there is no longer a restriction on the illuminator antenna size (within limits imposed by the size of the vehicle, ship, or aircraft carrying the illuminator). Most semi-active homing missiles home 'all-the-way', so that initially the missile is on its launcher alongside the illuminator, giving $R_1 = R_2$ (= R) for the purposes of acquisition. Recalculating the maximum acquisition range for the same seeker but taking advantage of the much greater ERP of a separate illuminator shows a marked increase. For example

P_I	500 W
G_I	47 dB
R_{max}	63 km

It is possible to extend the range still further by confining semi-active homing to the terminal phase. In this case $R_1 \neq R_2$. Assuming a terminal homing phase of 15 km (= R_2) gives R_1 = 260 km on acquisition. In practice, the figure would be rather less because of the greater atmospheric attenuation; nevertheless, long-range acquisition is certainly possible. On the other hand, active terminal homing, with the aid of some other, much-cheaper, method of mid-course guidance, can achieve the same result with greater autonomy; it is questionable therefore whether semi-active terminal homing is worth while.

3.4 Choice of wavelength

The wavelength is inevitably short, 30 mm or less, because of the need for a narrow antenna beam from an antenna confined within the small diameter of the missile. Even the fairly large early missiles rarely had antennas wider than 0.45 m; typical values nowadays are 0.3 m for a large missile, such as a medium range SAM or a sea-skimmer, 0.15 m for a short-range AAM, and even less, 0.1 m, for a battlefield anti-tank missile. However, the available size of antenna is not the only factor in the choice of wavelength; others are

> atmospheric and rain attenuation
> size and weight of components
> cost
> availability of transmitter power
> receiver noise performance

These factors will now be considered in turn.

Size of antenna

To achieve adequate resolution and tracking precision, a beamwidth of about 0.1 radian is the maximum. The beamwidth of an antenna of circular aperture is given by the approximate formula $1.5\lambda/d$, where d is the antenna diameter. This leads to the following table of maximum wavelength for various antenna diameters

d (m)	λ (mm)	Practical λ (mm)
0.3	20	20–22
0.15	10	8
0.1	6.7	3.2

Not all of the theoretically calculated wavelengths are suitable, as will emerge from the following section on atmospheric attenuation, and the third column gives the practical values.

Atmospheric and rain attenuation

This is a major factor at these short wavelengths. Graphs of atmospheric and rain attenuation appear in many standard works[2] but figure 3.1 is a sketch as a reminder. Consider the range equation examples appearing earlier (section 3.2). The active homer had an acquisition range of 12 km; with $\lambda = 8$ mm the clear weather attenuation is 0.05 dB km^{-1} (one way), giving a total loss of 1.2 dB, which is quite modest. In rain of intensity 4 mm h^{-1} extending the whole way, the one-way attenuation is 1.1 dB km^{-1} which corresponds to a total loss of

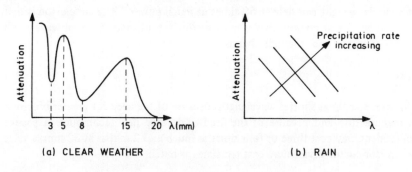

Figure 3.1   *Atmospheric and rain attenuation*

26.4 dB. To compensate for this by increasing transmitter power from 50 W to 22 kW is out of the question. If an increase of wavelength is unacceptable, then the acquisition range (with no increase of transmitter power) falls to 5.8 km. In practice, there would be a compromise between increase of transmitter power, and loss of acquisition range.

The semi-active homer of the example in section 3.3 had an acquisition range all-the-way, of 63 km; with λ = 22 mm the clear weather attenuation is 0.01 db km^{-1} one way, giving a total loss of 1.3 dB, corresponding to a small reduction of range to 58 km. In rain of 4 mm h^{-1}, the attenuation is 0.15 dB km^{-1}, representing a total loss of 19 dB. A corresponding increase of illuminating power is impracticable but a possible compromise would be a four-fold increase of power to 2 kW with a reduction of acquisition range to 35 km. Allowing for rain at these short wave lengths is costly; it is important, therefore, to specify no more allowance than is necessary. This means taking into account the conditions and the climate[3] in which the missile system is likely to be operating, and accepting some risk of mission failure as a result of this cause.

Size and weight of components

The size of RF components tends to increase with wavelength, though this is not inevitable. It is possible to use miniaturized components, such as microstrip and microwave integrated circuits (MIC) in semi-active seekers. On the other hand, active seekers still require bulky and heavy components such as the transmitter and its associated power supply, and standard hollow waveguide feeders. The dependence of the transmitter size and weight on wavelength is less marked than would be expected; the smaller size of valve at shorter wavelengths is often more than offset by the greater size and weight of ancillaries such as the magnet, cooling arrangements, and the high-voltage power supply. For example, a magnetron capable of delivering 3 kW peak power at 8 mm is similar in size and weight to one capable of delivering 10 times as much power at 32 mm[4], while a klystron of a similar weight can deliver 50 times as much power[5]. The conclusion is that size and weight of components is not a major factor in the choice of wavelength except at extremes of the radar bands.

Cost

This rises steeply at shorter wavelengths because of the need for greater precision in manufacture and because of the need to develop new techniques. A passive component may cost three or four times as much at 3.2 mm as at 32 mm; a valve or semiconductor device may cost ten times as much.

Availability of transmitter power

This is of more importance in semi-active systems, where other constraints are fewer. At 20 mm or longer there is unlikely to be a limit on what is available[6]; at 8 mm the upper limit for conventional power sources is about 1 kW mean, and at 3 mm it is about 100 W mean[7]. High-powered mm-wave sources, such as the gyrotron, are in existence but they are still in the experimental stage.

Receiver noise performance

The sensitivity of a microwave receiver depends mainly upon internal noise from RF stages. At wavelengths of 22 mm and longer, it is possible to incorporate low-noise amplifiers such as the FET to attain noise figures of 3 dB or 4 dB without cooling. Low-noise amplifiers are not available generally at shorter wavelengths, and reliance has to be placed on low-noise mixers. There has been much improvement recently so that receiver noise figures of 8 dB are attainable at 8 mm, or even at 3 mm. On the other hand, external noise[8] is not negligible at mm wavelengths, and this factor worsens the noise figure by a few dB.

Summary of the effects of wavelength

The preceding discussions are summarized in the qualitative graph of figure 3.2. From this it is possible to draw the following general conclusions about the choice of wavelength

semi-active, range up to 100 km	32 mm or 22 mm
active, range up to 15 km	22 mm or 8 mm
active, range of 1 km or 2 km	8 mm or 3 mm
radiometric, passive	8 mm or 3 mm

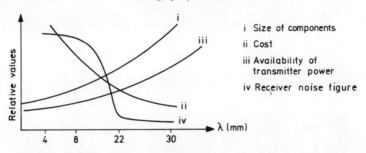

Figure 3.2   *Factors depending on wavelength*

References

1. Skolnik, M. I., *Introduction to Radar Systems*, 2nd edn, McGraw-Hill, 1980, Chapter 2.
2. Skolnik, M. I., *Introduction to Radar Systems*, 2nd edn, McGraw-Hill, 1980, Section 13.7.
3. *CCIR XIIIth Plenary Assembly Report*, Vol. 5, ITU, Geneva, 1974.
4. English Electric Valve Co., advertisement in *Microwave Journal*, June 1982, p. 69.
5. Thomson–CSF, 'International Market Place', *Microwave Journal*, November 1982, p. 34H.
6. Kaiser, F. L., 'Microwave Technology Review', *Microwave Journal*, July 1977, p. 23.
7. Ewell, G. L., *et al.*, 'High power millimetre wave radar transmitters', *Microwave Journal*, August 1980, p. 57.
8. Skolnik, M. I., *Introduction to Radar Systems*, 2nd edn, McGraw-Hill, 1980, Section 12.8.

4

Seeker Antenna Systems

4.1 Introduction

The purpose of the antenna is to sense the direction of radiation from the target by means of radar tracking techniques, thus defining the sight line continuously. A pair of rate gyroscopes mounted orthogonally on the gimbals senses sight line rate with respect to space axes in each plane, providing the information for proportional navigation. Angle tracking methods already in use or under development fall into the following categories

Amplitude comparison; monopulse; sequential (conical scan)
Phase comparison
Interferometer
Electronic beam steering (the phased array)

A brief account of the principles of tracking radar, illustrated by amplitude comparison methods, appears in section 1.3; the purpose of the present chapter is to describe the application of these principles to radar homing.

4.2 Amplitude comparison[1]

Conical scan

This is a well-established technique used mainly in older systems such as Thunderbird II, Bloodhound, and Hawk; a single-plane oscillating beam is used in sea-skimmers. Figures 4.1 and 4.2 are photographs of a two-plane and a single-plane antenna respectively. The principles of conical scan have been described

Figure 4.1   *Two-plane seeker antenna (courtesy of Royal Military College of Science)*

Figure 4.2   *Single-plane seeker antenna (courtesy of Royal Military College of Science)*

already in chapter 1; figure 4.3 is a block diagram of a guidance receiver using this technique. The output of the envelope detector is a DC voltage which varies at conical scan rate in accordance with the variation in signal strength due to pointing error; a pair of reference alternating voltages derived from a generator keyed to the antenna spinner shaft allows the two phase detectors to separate

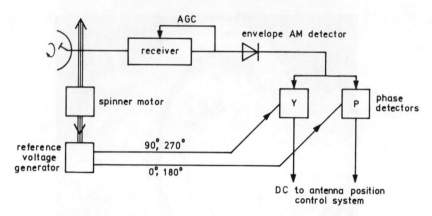

Figure 4.3   *Conical scan seeker*

pitch (elevation) and yaw (azimuth) errors, producing a pair of DC voltages proportional in magnitude and sign to the respective errors. These voltages are used to energize their respective antenna position control systems as described in section 1.2. AGC applied to the receiver ensures that the system responds to the error only and is not influenced by the mean signal strength. A range gate or a velocity gate ensures that the system tracks only the selected target echo.

Conical scan is not favoured nowadays, except for very short-range, simple guidance systems, because of its vulnerability to a form of ECM known as 'spin frequency jamming' in which the jammer re-radiates the seeker transmission, amplitude modulated at the conical scan frequency, thus creating spurious angle error signals. A semi-active seeker is less vulnerable to this effect than is an active seeker because it does not disclose its conical scan frequency by a transmission; nevertheless an enemy would probably be aware that a certain missile used conical scan and would have a notion of the spin rate, thus enabling him to deploy effective ECM if he believed he was under attack from this type of missile.

Monopulse

Amplitude comparison monopulse trackers as described in section 1.3 have not been popular in seekers because of the elaborate bulky antenna feed system and the multi-channel receiver required. Phase comparison monopulse, described in the next section, has a less cumbersome antenna feed, while recent developments in integrated circuit technology have made it possible to incorporate the multi-channel receiver within the limited space of a missile.

Cassegrain feed[2]

Amplitude comparison antennas suffer from blockage of the main aperture by the primary source situated in the focal plane and by the associated feeders; this is especially serious in a seeker, where the antenna aperture is already limited severely by the small diameter of the missile. J. Cassegrain (1672) solved this difficulty in the optical reflecting telescope (figure 4.4a) by placing a small hyperbolic sub-reflector M between the pole of the main reflector and the focus S, so that it intercepted rays converging on the focus, reflecting them through a small hole in the pole of the mirror to a new focus S' just behind the pole. In the radar application, blockage of the aperture by the sub-reflector is avoided by a clever use of polarization. The sub-reflector, which can be plane without serious loss of performance, consists of a set of closely spaced parallel wires perpendicular to the plane of polarization of the incident plane wave V (figure 4.4b), hence it is almost transparent. On the surface of the main reflector is a similar set of wires, oriented at 45° to those on the sub-reflector (figure 4.4c). The wires resolve the incident wave V into two equal components — one perpendicular to the wires (A) and one parallel to the wires (B). B is reflected at the surface, while A passes through the λ/4 thick foam dielectric sheet forming the body of the mirror, is reflected at the metal backing, and emerges as A', in anti-phase to A. A' recombines with B to form a horizontally polarized wave H converging on the focus S. The horizontal wires of M reflect H through the hole at the pole of the main reflector on to the primary source S', which is oriented for horizontal polarization. Aperture blockage by the hole in the mirror is considerably less and more regular than that due to conventional front feed, resulting in higher gain and smaller side-lobes.

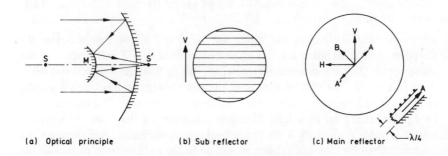

(a) Optical principle   (b) Sub reflector   (c) Main reflector

Figure 4.4   *Cassegrain antenna*

4.3 Phase comparison monopulse[3]

The amplitude comparison antenna is a curved reflector, occupying a considerable depth, and hence it sweeps out a considerable volume when it swings over its arc, which may be ± 45° or more. To allow for this movement, the mirror diameter must be not more than about 75 per cent of the internal diameter of the missile. A phase comparison antenna and associated feeders can be constructed upon a thin flat plate, which requires less space in which to turn, and hence its diameter can be a greater proportion of the missile diameter, say 85 per cent, giving a correspondingly better angular precision. The basic principle of the phase comparison antenna is that of the interferometer (figure 4.5), consisting of a pair of antennas, A and B, spaced a distance d ($\gg \lambda$); the axis is defined as

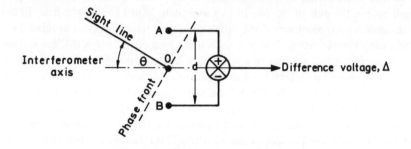

Figure 4.5   *Principle of the interferometer*

the line through the mid-point, O, perpendicular to AB. The phase front of the signal from a distant target lying in direction θ reaches A before it reaches O, hence the signal voltage at A can be represented by $\exp(+j\phi)$, where $\phi = 2\pi/\lambda \times (d \sin \theta)/2$; similarly the voltage at B is $\exp(-j\phi)$. Subtracting these voltages yields a difference voltage

$$\Delta = 2 \sin\left[\frac{\pi d}{\lambda} \sin \theta\right] \exp(j\phi/2) \tag{4.1}$$

whose magnitude contains the magnitude of θ and whose sign indicates the direction. Note that there is a fixed phase difference of $\pi/2$ with respect to the phase centre O of the interferometer, which has to be taken into account when demodulating Δ. The moving version of the phase comparison antenna is usually an array of $\lambda/2$ elements on a circular flat plate (figure 4.6). The elements may be dipoles etched on to a light dielectric substrate, or they may be slots cut into a ground plane with a dielectric backing; furthermore, the elements are interconnected by strip line feeders etched on to the back or front surface of the plate, as appropriate. Much work has been done recently on this type of antenna and there is quite an extensive literature available — see ref. 4, for example. For phase comparison in two planes, the array is divided into four quadrants as shown in figure 4.6. Adding the voltages from all four quadrants in a hybrid yields the sum signal, Σ. Adding upper pairs and adding lower pairs, and then subtracting the sums, yields the pitch error voltage Δp; similarly, comparing right-hand and left-hand pairs yields the azimuth or yaw error voltage Δy; each error voltage drives its own antenna position control system. An analysis is as follows. Each pair of quadrants is equivalent to one antenna of an interferometer pair; the phase centre of the upper pair is at U, the centroid of their area; that of the lower pair is at D, giving an interferometer base line length of d' (figure 4.7).

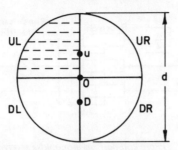

Figure 4.6   *Flat plate array*

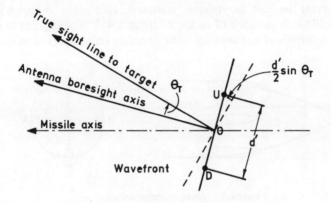

Figure 4.7   *Phase comparison antenna*

A pointing error θ_T gives rise to an error voltage proportional to $\sin(\pi d' \sin \theta_T/\lambda)$; since θ_T is small, a few milliradians, the voltage is directly proportional in magnitude and sign to θ_T. The expression for the error voltage in the other plane is similar. The error voltages are amplified and demodulated in a monopulse receiver, as shown in figure 4.8; the purpose of the $\pi/2$ phase shifter is to compensate for the $\pi/2$ phase shift that occurs when the difference is taken. The phase comparison antenna features in many modern missiles, such as Patriot and Sky Flash, and also in terminally guided sub-munitions, where space is especially at a premium. Figure 4.1 shows a photograph of a typical antenna.

4.4 Interferometer systems

The fixed interferometer has the advantage that it makes use of the outside diameter of the missile, with the possibility of greater angular precision. Interferometers have been used in several US Navy SAMs, though the most recent example is in the British Sea Dart missile. An interferometer antenna for two-

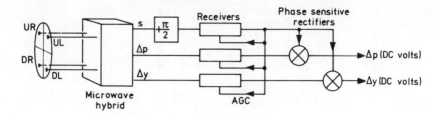

Figure 4.8   *Phase comparison monopulse receiver*

plane homing consists of two pairs of individual antennas fixed around the perimeter of the missile (figure 4.9). The antennas are end-fire, such as polyrods, so as to present the least aerodynamic resistance; such antennas have a modest gain (7–10 dB) with coverage of an arc of about ± 45°. As explained in section 4.3, the interferometer measures sight-line direction with respect to its own axis,

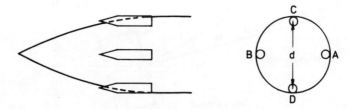

Figure 4.9   *Interferometer antennas*

which in this case is fixed with respect to the missile. There is, therefore, no direct way of measuring the rate of change of sight line with respect to inertial axes, as required for proportional navigation. To overcome this difficulty the guidance system incorporates a line-of-sight (LOS) platform, gimballed in both planes, to define the LOS mechanically; two rate gyroscopes, mounted ortho- gonally on the platform, give the required sight-line rates. The platform is oriented initially by commands from the parent target tracker while the missile is still on the launcher, and it is stabilized in this direction during the boost phase by the rate gyroscopes. Once guidance commences, it is steered by signals from the interferometers as follows: the two antennas of an interferometer pair generate voltages $\exp(+j\phi_s)$ and $\exp(-j\phi_s)$ respectively, where ϕ_s is the phase angle corresponding to the direction of the sight line with respect to the missile axis (figure 4.10a). ϕ_s is compared with a synthetic phase angle, ϕ_D, derived from the orientation θ_D of the platform with respect to the missile axis (figure 4.10b). The output of the phase comparator is a difference voltage given by

$$\Delta = 2j\sin(\phi_s - \phi_D) \text{ (compare equation 4.1)}$$

$$= 2j\sin\left[\frac{\pi d}{\lambda}\left(\sin\theta_s - \sin\theta_D\right)\right] \tag{4.2}$$

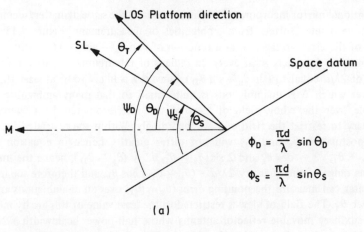

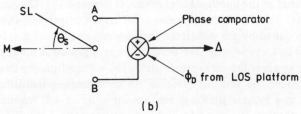

Figure 4.10 *Interferometer system*

$$= 2j \sin\left[\frac{2\pi d}{\lambda} \cos\left(\frac{\theta_s + \theta_D}{2}\right) \sin\left(\frac{\theta_s - \theta_D}{2}\right)\right] \qquad (4.3)$$

Since the pointing error θ_T is small, $(\theta_s + \theta_D)/2 \approx \theta_s$ and $2\sin(\theta_s - \theta_D/2 \approx \theta_T$. Hence

$$\Delta = 2j \sin\left[\frac{\pi d}{\lambda} (\cos \theta_s) \theta_T\right]$$

Provided that the system is tracking correctly, θ_T is small enough for this expression to be simplified further

$$\Delta = 2j \frac{\pi d}{\lambda} (\cos \theta_s) \theta_T \qquad (4.4)$$

giving Δ proportional to θ_T. Note that Δ is also proportional to $\cos \theta_s$; the consequent limitation of the field of view is expressed by the factor $\{G(\theta_s)\}^{\frac{1}{2}} \times \cos \theta_s$, where $G(\theta_s)$ is the directivity factor of each interferometer antenna. Hence the practical field of view is about 45° either side of the interferometer axis. There are two disadvantages to this simple interferometer: firstly, the gain of the antennas is likely to be considerably less than that of a conventional mirror incorporated within a missile of the same diameter; secondly, an interferometer suffers from ambiguities of measurement. Figure 4.11 is a graph of the error voltage, Δ, as a function of $(\sin\theta_s - \sin\theta_D)$ (equation 4.2), showing that Δ repeats itself every 2π radians of the argument — that is, over a range of $2\lambda/d$ equal to $(\sin\theta_s - \sin\theta_D)$. Since d/λ is likely to be at least 10, the arc over which Δ is unambiguous is quite small so that many ambiguities are possible over the whole field of view of the interferometer. It is therefore necessary to restrict the field of view accordingly. Within the restricted field of view postulated, θ_s and θ_D will not differ greatly, hence in equation 4.3, $\cos(\theta_s + \theta_D)/2 \approx \cos\theta_s$ and $2\sin(\theta_s - \theta_D)/2 \approx (\theta_s - \theta_D)$, hence the unambiguous range of the quantity $2\lambda/d = (\theta_s - \theta_D)\cos\theta_s$, and therefore the interferometer can measure the pointing error $(\theta_s - \theta_D)$ over an unambiguous arc of $2\lambda/d \sec\theta_s$. The field of view is restricted to the least value of this arc by means of an auxiliary movable reflector antenna whose half-power beamwidth is $2\lambda/d$, and which is pivoted at the interferometer centre, O (figure 4.12). The auxiliary antenna is oriented initially in the direction in which the target lies. Once the target has been acquired by the seeker, the auxiliary antenna is caused to follow the LOS platform by a servo system; this need not be very precise as the auxiliary antenna has only to cover the arc approximately. The voltage from the auxiliary antenna also provides the sum or reference for demodulating the difference channel voltages in a manner similar to that shown in figure 4.8. Without this reference voltage there is no output from the phase detectors of the difference channels, hence signals from outside the field of view of the auxiliary antenna are ineffective. This may seem to be an unduly complicated arrangement, but section 4.5 shows that it can provide a worthwhile improvement in angular precision, especially where space for a conventional antenna is limited.

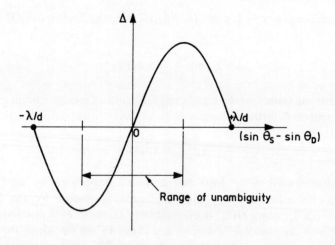

Figure 4.11 *Interferometer ambiguities*

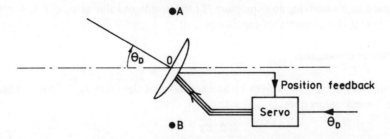

Figure 4.12 *Auxiliary mirror for interferometer*

4.5 Phased arrays

Rapid inertialess scanning of the antenna beam by a phased array[5] is an attractive concept which already finds widespread application in radar. Its advantage for missile seekers is that it does away with moving parts, so reducing demands on space and power. It can also make better use of the diameter of the missile.

The most straightforward form is a flat plate fixed perpendicularly to the missile axis and protected by a conventional radome. A more novel approach is to site the elements on the surface of the nose of the missile, thus eliminating the radome; such an array is known as 'conformal'. Several workers have been studying the application of conformal arrays to missile seekers, and Mitchell[6] has given a critical review together with an extensive list of references. The conformal array has, however, the disadvantages that some elements are always blind, creating an asymmetrical radiation pattern with high side-lobes; much of the present effort is directed towards reducing these side-lobes.

A difficulty with all fixed arrays is that there is no direct way of measuring sight-line rate with respect to inertial axes. $\dot{\psi}_s$ (section 1.2) is obtainable by measuring sight-line rate with respect to missile axes and adding it to missile body rate, $\dot{\psi}_m$, with respect to inertial axes. But this arrangement is potentially unstable because the required quantity, $\dot{\psi}_m$, is derived from itself, though damping in the aerodynamic loop can make the system sufficiently stable provided that the rates are slow. The most likely applications are, therefore, against slow-moving targets such as ships or land vehicles, or against stationary targets such as radar installations.

4.6 Angle tracking precision – a comparison

There has already been mention of tracking precision as a feature of an angle tracking system. This is a useful criterion for comparing the performance of different types of seeker. The measure of precision is the standard deviation, σ_θ of random tracking error due to noise. The purpose of this section is to make a comparison by deriving an expression for the optimum value of σ_θ for each type of tracker under comparable conditions.

Amplitude comparison

Barton[7] shows that σ_θ is given by an expression of the form $\theta_3/k(S{:}N)^{\frac{1}{2}}$, where k is a dimensionless number defined as

$$\frac{d(\Delta/\Sigma)}{d(\theta_T/\theta_3)}\bigg|_{\theta_T = 0} \tag{4.5}$$

where Δ is the error voltage or output of the difference channel of the receiver arising from a misalignment θ_T of the seeker bore-sight axis from the sight line to the target, Δ is expressed as a proportion of the mean amplitude, Σ, of the target signal at the output of the sum channel, and θ_3 is the half-power beam-width of the antenna. The normalizations ensure that the expression is general, depending only upon the configuration of the antenna. Barton gives $k = 1.5$ for monopulse; $k = 1$ for conical scan, one-way as in semi-active homing, and slightly more for two-way. For convenience, $k = 1$ will be adopted for conical scan. Taking θ_3 as $1.5\ \lambda/d$ for a circular aperture of diameter d gives

$$\sigma_\theta = \frac{\lambda}{d}\ \frac{1}{(S{:}N)^{\frac{1}{2}}}\quad\text{monopulse} \tag{4.6}$$

$$\sigma_\theta = \frac{1.5\lambda}{d}\ \frac{1}{(S{:}N)^{\frac{1}{2}}}\quad\text{conical scan} \tag{4.7}$$

Phase comparison

For the system described above in section 4.3, it can be shown that

$$k = \frac{4}{3}\ \frac{d}{\lambda}\ \theta_3\quad\text{(see appendix A)}$$

hence

$$\sigma_\theta = 0.75\ \frac{\lambda}{d}\ \frac{1}{(S{:}N)^{\frac{1}{2}}}$$

The derivation of this expression requires no assumption about the relationship between θ_3 and λ/d, but the antenna gain must remain the same as in amplitude comparison, so that $S{:}N$ is the same in all cases.

Interferometer with auxiliary antenna

Appendix A also shows that

$$\frac{d(\Delta/\Sigma)}{d\theta_T}\bigg|_{\theta_T = 0} = \frac{\pi\,d_m}{\lambda}\left(\frac{2G_I}{G_D}\right)^{\frac{1}{2}}\times\cos\theta_s \tag{4.8}$$

for this system, where G_I is the gain of each interferometer antenna and G_D is the gain of the auxiliary 'dish' antenna, whence

$$\sigma_\theta = \frac{\lambda}{d_m} \left[\frac{1}{\pi \sqrt{2}} \left(\frac{G_D}{G_I} \right)^{\frac{1}{2}} \frac{1}{\cos \theta_s} \right] \frac{1}{(S:N)^{\frac{1}{2}}} \qquad (4.9)$$

There is no basis for direct comparison with amplitude and phase comparison systems, as it is necessary to specify a value for $G_D:G_I$. It is also necessary to stipulate that G_D is the same as the gain of the antenna in each of the moving antenna systems; if it is not, $S:N$ must be scaled accordingly.

Phased array (flat plate)

The expression for σ_θ is as for a moving phase comparison antenna, but includes the factor (sec θ_s) to allow for the antenna being fixed.

Summary

Table 4.1 summarizes the results of this section.

The relative merits of the systems are obvious, except for that of the interferometer with auxiliary antenna. Let us take a specific example: each antenna has, say, a gain of 10 dB. As stated in section 4.4, the auxiliary antenna requires a half-power beamwidth of $2\lambda/d_m$, which corresponds approximately to a gain of $[\pi d_m/2\lambda]^2$. These figures give a relative error of 0.11 sec θ_s, which at the extreme value of $\theta_s = 45°$ is 0.16. The relative errors of the other systems are in terms of λ/d, where d is the antenna diameter. The interferometer with auxiliary antenna will have the best performance of all if $d/\lambda < 4.7$. This could be the case in a small missile, or where the space for a conventional antenna is restricted by other requirements.

Table 4.1 Comparison of precision of seeker antenna systems

Type of system	Relative error $(\sigma_\theta \times (S/N)^{\frac{1}{2}})$
Amplitude comparison, monopulse	λ/d
Amplitude comparison, conical scan	$1.5\ (\lambda/d)$
Phase comparison	$0.75\ (\lambda/d)$
Interferometer with auxiliary antenna	$0.23\ (\lambda/d_m) \times (G_D/G_I)^{\frac{1}{2}} \times \sec \theta_s$
Phased array (flat plate)	$0.75\ \dfrac{\lambda}{d} \times \sec \theta_s$

4.7 Multiplexing

A true two-plane monopulse receiver requires three separate channels — one for the sum Σ, and one for each difference signal Δp and Δy. In the confined space of a missile it may not be possible to accommodate all three, so that it becomes necessary to multiplex the signals through two channels or even through just one. The cost is partial loss of the monopulse feature, and some loss of signal strength in the difference channels as a result of time sharing. Figure 4.13 is the block diagram of a possible two-channel receiver; Δp and Δy are multiplexed in time by switch waves at a rate of a few hundred hertz and combined with Σ so that one channel carries Σ + multiplexed Δ, the other Σ — multiplexed Δ. The presence of Σ ensures a high signal level in each channel, thereby avoiding the non-linearities which tend to occur at low levels. It is, however, essential to control the relative gains and phase shifts of the two channels in order to avoid false interpretations arising from differences in these two constants.

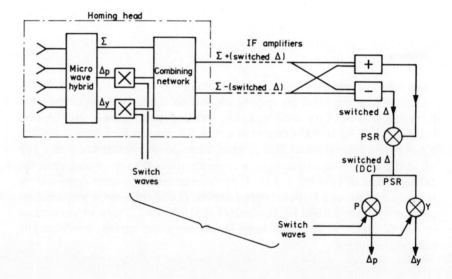

Figure 4.13 *Multiplexing into two channels*

4.8 Radomes[8,9]

The radome is the transparent nose of the missile which protects the antenna of the seeker, while preserving the aerodynamic shape of the missile. References to missile radome technology are scarce, and hence this section gives a brief account of the technology in general, as well as dealing in more detail with the electromagnetic properties.

The radome, occupying a most vulnerable position in flight, is subject to particularly harsh conditions; ideally it should meet the following requirements

Be transparent to EM waves at the working frequency
Introduce no distortion of the wavefront (aberration)
Be of good aerodynamic shape
Be resistant to shock and vibration
Be resistant to aerodynamic heating
Be resistant to erosion by rain
Be light
Be cheap and easy to manufacture

Radome materials in common use, and their general properties, will now be discussed.

Fibre-reinforced organic resins

The fibre is usually glass; a variety of thermo-setting resin types, such as phenolic, polyester, and epoxy, is available. Radomes constructed of these are relatively cheap and easy to make, even in quite large sizes.
Typical physical properties are

Relative permittivity, ϵ_r	4-6
Loss tangent, tan δ	0.01
Density, figures not available, but light	
Ultimate tensile strength (UTS)	40 k psi
Compressive strength (CS)	$20\text{-}40 \times 10^3$ psi
Flexural strength (FS)	$30\text{-}50 \times 10^3$ psi
Maximum temperature	$260°C$; that is, only suitable for missile speeds of less than M1.5-M2

Poor resistance to rain erosion
Control of thickness and hence of aberration is not easy; surface has to be pared down afterwards

Refractory oxides

Alumina is the most used, others are magnesia, fused silica etc. Radomes of these materials are expensive because of the high temperature needed in forming; the process and material are more suitable for small radomes.
Physical properties of alumina are

ϵ_r, at microwaves	8-9
tan δ at microwaves	0.002-0.0001
Density	220 lb per ft^3
UTS	27×10^3 psi
CS	280×10^3 psi

FS 30×10^3 psi
Maximum temperature $1700°C$ — almost no limit on missile
                                            speed
Wall thickness, can be controlled to 0.002 inch, hence better control of aberration
Good resistance to rain erosion

Glass ceramics

Typical of this class of material are Pyroceram and glass-bonded mica. Again
radomes are expensive to construct, and it is difficult to make large ones. Typical
physical properties are

ϵ_r, at microwaves   5.5
$\tan \delta$ 0.0003 (Pyroceram)
                                            0.003 (glass-bonded mica)
Density   160 lb per ft^3
UTS 7×10^3 psi $\Big\}$ glass-bonded mica
CS 35×10^3 psi
Maximum temperature $500°C$
Better resistance to thermal shock than alumina
Good resistance to rain erosion

Silicon nitride

This is a newer material. Physical properties are

ϵ_r 5.5
$\tan \delta$ 0.001
Good resistance to thermal shock

Electromagnetic properties

The most obvious property is the microwave attenuation. Fortunately, low
attenuation is quite easy to attain; even the most lossy materials, the fibre-
reinforced organic resins, introduce attenuation of no more than a few tenths of
a decibel. Aberration is more of a problem; it is of consequence because it can
lead to instability in the missile flight. Figure 4.14 explains aberration and defines
the terms. The sight line, making an angle of θ_s with the missile axis, is bent
through an angle ϵ so that it arrives at point O as if it had come from a direction
θ_a. The magnitude and sign of ϵ is not of prime importance; what does matter
is the slope of the aberration characteristic, or the rate of change of ϵ with θ_s,
because proportional navigation depends upon the rate of change of sight-line
direction.

Figure 4.15 shows an idealized characteristic with a uniform positive slope
(a), one with a uniform negative slope (b), and one with a varying slope (c); note

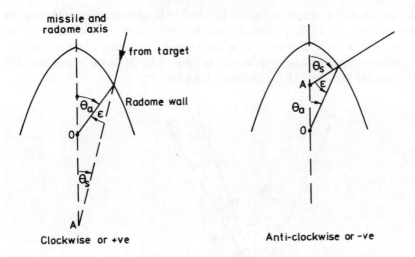

Figure 4.14 *Radome aberration*

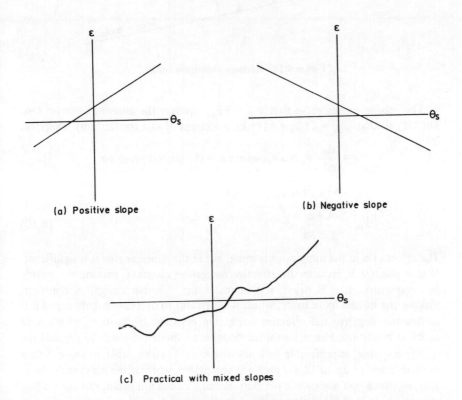

Figure 4.15 *Radome aberration characteristics*

that the aberration is not necessarily zero at $\theta_s = 0$. Aberration is quite small; its magnitude is only a few milliradians and its slope is only a few per cent. Nevertheless, such a slope is enough to be troublesome, particularly if it is negative, as the following simple analysis shows. In figure 4.16 OM is the missile axis, OT the true sight line, and OA the aberrated sight line.

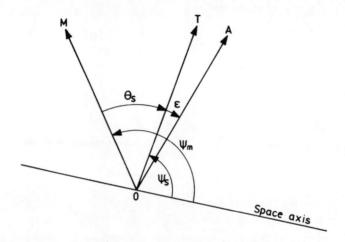

Figure 4.16 *Radome aberration angles*

The missile navigates so that $\dot{\psi}_m = \lambda \dot{\psi}_s$, however the antenna observes OA, not OT, so that $\dot{\psi}_m = \lambda (\dot{\psi}_s + \dot{\epsilon})$ (taking account of sign convention). Therefore

$$\dot{\epsilon} = \frac{d\epsilon}{d\theta_s} \dot{\theta}_s = \alpha \dot{\theta}_s, \text{ where } \alpha \text{ is the aberration slope}$$

$$\dot{\psi}_m = \lambda (\dot{\psi}_s + \alpha \dot{\psi}_s - \alpha \dot{\psi}_m)$$

$$\dot{\psi}_m = \frac{\lambda + \alpha}{1 + \lambda \alpha} \dot{\psi}_s \qquad (4.10)$$

The effect of α in the numerator is small, but in the denominator it is significant. If α is positive it reduces the effective navigation constant, making the missile less responsive. If it is negative it increases the effective navigation constant, making the missile more lively, which is equivalent to decreasing damping; if it is sufficiently negative the effective navigation constant tends to ∞, which is an unstable condition. Hence a positive slope, even though it tends to degrade the missile response, is preferable to a negative one. As an example, in one missile a positive slope of up to 12 per cent was acceptable, whereas the maximum allowable negative slope was only 2 per cent. Because of the importance of controlling aberration, it is the practice to measure the aberration characteristic of every radome in both principal planes. Faulty radomes can be corrected by shaving off layers of the inside wall.

The principal causes of radome aberration are the inner and outer walls not being parallel, and inhomogeneity of the radome material. Radome aberration is worst at glancing incidence, hence the best shape is hemispherical and the worst is a finely pointed cone or ogive. Fibre-reinforced organic resin is a poor material for construction as the radome is built up layer by layer, making it difficult to maintain constant thickness and homogeneity; fortunately, radomes of this material are usually to be found on slow missiles where it is possible to employ a blunt shape which is more favourable. Fast missiles require a finely pointed radome, but the radomes are cast, giving better control of thickness and homogeneity.

Conclusions

There are many conflicting factors in the choice of material and in the design of a radome, as the list above demonstrates. Fibre-reinforced organic resins are still used for subsonic missiles, as they are cheap; though even here, glass ceramics with their superior and more consistent electrical properties, are preferred for smaller missiles. Medium-speed missiles (M1-M2.5) tend to use glass ceramics. High-speed missiles (M3 +) require refractory oxides, though silicon nitride is a useful alternative because of its better resistance to thermal shock.

References

1. Skolnik, M. I., *Introduction to Radar Systems*, 2nd edn, McGraw-Hill, 1980, Sections 5.1-5.4.
2. Skolnik, M. I., *Introduction to Radar Systems*, 2nd edn, McGraw-Hill, 1980, Section 7.5.
3. Skolnik, M. I., *Introduction to Radar Systems*, 2nd edn, McGraw-Hill, 1980, Section 5.4.
4. James, J. R., Hall, P. S. and Wood, C., 'Microstrip antenna theory and design', *IEE Electromagnetic Waves Series*, No. 12, 1981.
5. Skolnik, M. I., *Introduction to Radar Systems*, 2nd edn, McGraw-Hill, 1980, Chapter 8.
6. Mitchell, P. J., 'Conformal arrays for guided weapons; a review', *Military Microwaves*, 1980, pp. 457-69.
7. Barton, D. K., 'Detection and Measurement', Chapter 2 in Brookner, E. (ed.), *Radar Technology*, Artech House, 1971.
8. Fudge, G. L. and Moore, T. S., 'Ceramic radome development for g.w.', *2nd Military Microwave Conference Proceedings 1980*, Microwave Exhibitions and Publications (UK) Ltd, 1980, pp. 584-8.
9. Rudge, A. W., *et al., 2nd Military Microwave Conference Proceedings 1980*, Microwave Exhibitions and Publications (UK) Ltd, 1980, p. 555.

5

Principles of Doppler Homing Guidance

5.1 Introduction

The seeker in a Doppler system makes use of the differential Doppler frequency to distinguish small but fast-moving airborne targets from each other and from any background of terrain and precipitation clutter. Early Doppler homers used pure CW illumination of the target, but recently the tendency has been to employ some type of modulation such as pulse or FMCW, leading to pulse-Doppler systems etc. However, whatever the nature of the modulation, the principles of Doppler signal processing are the same. The Doppler frequency in simple engagement configurations has been derived in chapter 2. In practice, configurations will vary widely, so that possible Doppler frequencies may lie anywhere within a broad band; the purpose of this chapter is to derive general expressions for the Doppler frequency in both active and semi-active homing, and to illustrate these expressions by specific examples.

5.2 Doppler frequency spectra

Active homing

Figure 5.1 shows the general engagement geometry for an active homer. The antenna is pointing at the target T; the Doppler shift on the target echo is

$$f_D = 2 (u_m \cos \delta + u_t \cos \alpha)/\lambda \qquad (5.1)$$

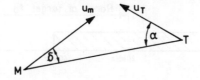

Figure 5.1   *Active homing – target geometry*

Since α and δ almost certainly change during an engagement, f_D changes and the velocity gate must track it. The maximum value of f_D is when the missile, M, and T are flying towards each other and is given by

$$f_{D_{max}} = 2 \, (u_m + u_t)/\lambda \qquad (5.2)$$

and conversely the minimum value is $2 \, (u_m - u_t)/\lambda$. To give an idea of the magnitude, take $\lambda = 30$ mm, $u_m = 600$ m s^{-1}, $u_t = 300$ m s^{-1}

$$f_{D_{max}} = 60 \text{ kHz}; \quad f_{D_{min}} = 20 \text{ kHz}$$

If the target is at the crossing point ($\alpha = 90°$) and the missile is flying directly towards it, $f_D = 40$ kHz. Hence, in this example the target Doppler frequency can lie anywhere between 20 and 60 kHz. Consider now the effect of surface clutter. For simplicity, assume that the missile is flying horizontally and that the seeker antenna is looking downwards at an angle δ, as shown in figure 5.2. The main beam strikes the ground at A, hence the Doppler frequency of the main

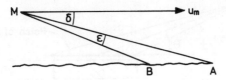

Figure 5.2   *Active homing geometry – surface clutter*

beam clutter (known as the ground spike) is $(2u_m/\lambda) \cos \delta$. However, the side-lobes of the antenna also give rise to clutter, which has a Doppler frequency of $(2u_m/\lambda)\cos(\delta + \epsilon)$ in the direction ϵ. Most antennas have significant side-lobes out to 90° but, since only those which look downwards contribute to clutter, the effective values of ϵ lie between $-\delta$ and $90° + \delta$. The clutter spectrum extends therefore from $2u_m/\lambda$ to $- (2u_m/\lambda) \sin \delta$. Using the same numerical data, the maximum value of the clutter Doppler frequency is 40 kHz; the minimum is at least zero and can be negative if the antenna has significant backwards-looking side-lobes. Figure 5.3 shows the extent of the clutter Doppler frequency spectrum, compared with the range of possible target Doppler frequencies, revealing that there is an overlap. Since it may well be necessary to engage a target whose

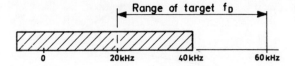

Figure 5.3   *Active homing – Doppler frequency spectra*

Doppler frequency falls within the clutter spectrum (target beyond the crossing point), it is necessary to estimate the clutter:target signal strength ratio (C:S). Calculations of clutter magnitude are invariably imprecise because the exact nature of the terrain is rarely known and because quantitative information about clutter is still incomplete; nevertheless, simple calculations can give an estimate. The following analysis follows Skolnik[1] ; figure 5.4 shows the geometry for clutter in the main beam, which is assumed to be of square cross-section and of angular width θ_3, the half-power beamwidth. Provided that $\theta_3 \ll \theta$, the grazing angle, the dimensions of the patch of clutter illuminated are as shown. Assign a value σ_0 to the radar cross-section of unit area of terrain, giving the total radar cross-section of the patch as $\sigma_0 \ (R\theta_3)^2 \times \operatorname{cosec} \theta$. If the terrain scatters diffusely, as do most terrains at short radar wavelengths, the scattering cross-section is independent of θ, by Lambert's Law of diffuse scattering, so that $\sigma_0 \operatorname{cosec} \theta$ is a constant, $= \sigma'$. Hence the radar cross-section of the clutter can be written as $\sigma' \ (R\theta_3)^2$.

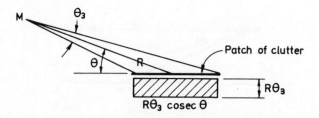

Figure 5.4   *Active homing – surface clutter*

The target lies in the main beam and, assuming that it is flying close to the surface of the earth, it is at approximately the same range as the patch of main lobe clutter. Hence C:S $= \sigma'(R\theta_3)^2/\sigma$, where σ is the radar cross-section of the target. The worst case is when the target is at maximum range R_{max}, where its echo is at a minimum, S_{min}, so that

$$C{:}S_{min} = \frac{\sigma'(R_{max} \ \theta_3)^2}{\sigma}$$

To give an idea of the magnitudes, take $R_{max} = 10$ km, $\theta_3 = 80$ mrad and $\sigma = 5$ m². The value of σ' depends upon terrain and frequency; various sources[2,3] give numerical data, and a typical value for land is -20 dB at 10 GHz. Substituting these figures in the formula gives $C{:}S_{min}$ = +30 dB. Over the sea the clutter level, which depends upon the sea state, tends to be lower but would still be troublesome. It is possible to derive an exact expression for the side-lobe clutter level[4] but it suffices here to assume that it follows the pattern of the side-lobes. A typical side-lobe level is -40 dB (two-way) with respect to the main lobe, giving a graph of $C{:}S_{min}$ against Doppler frequency of the form shown in figure 5.5. This shows that occasions may arise when the clutter obscures the target echo. To avoid this, it may be necessary to employ range discrimination as well as velocity discrimination, using a ranging modulation such as FMCW or pulse.

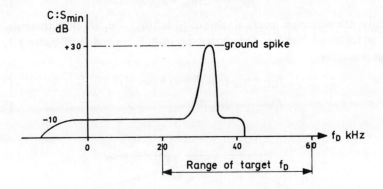

Figure 5.5   *Active homing – clutter levels*

Semi-active homing

Figure 5.6 shows the general geometry. Let the illuminator carrier frequency be f_c. The target receives a frequency $f_c + (u_t/\lambda) \cos \beta$ due to the movement of T with respect to I; this is re-radiated towards M with a further Doppler shift of

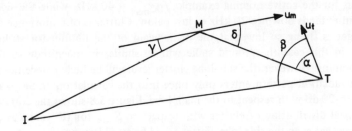

Figure 5.6   *Semi-active homing – target geometry*

$u_t \cos \alpha/\lambda$ due to the movement of M towards T; hence the seeker receives a signal frequency

$$f_c + [u_m \cos \delta + u_t (\cos \alpha + \cos \beta)]/\lambda$$

The guidance receiver extracts target f_D by comparing this frequency with that of the rear reference, which is $f_c - (u_m/\lambda) \cos \gamma$ due to movement of M away from I. Hence

$$f_D = [u_m (\cos \gamma + \cos \delta) + u_t (\cos \alpha + \cos \beta)]/\lambda \qquad (5.3)$$

The maximum value of f_D occurs when I, M, and T are collinear and M and T are flying towards each other, giving

$$f_{D_{max}} = 2(u_m + u_t)/\lambda \qquad (5.4)$$

Similarly, the minimum practical value of f_D is $2(u_m - u_t)/\lambda$. Both these are the same as for active homing. The clutter spectrum is derived from figure 5.7; by similar reasoning

$$f_{DC} = u_m [\cos (\delta + \epsilon) + \cos \gamma]/\lambda \qquad (5.5)$$

$$f_{DC_{max}} = 2u_m/\lambda \qquad (5.6)$$

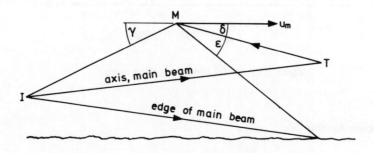

Figure 5.7   *Semi-active homing – clutter geometry*

Assuming γ is small, the minimum is $u_m (1 - \cos \delta)/\lambda$. Using the same numerical values as for the active homing example, $f_{DC_{max}} = 40$ kHz, while the side-lobe spectrum extends down to 20 kHz or just below. Clutter is of importance only if the target is flying so low that the main beam of the illuminator strikes the ground; in this case the ground spike will be similar in magnitude to that in active homing. However, the side-lobe clutter level will be higher since the seeker antenna radiation pattern enters only once into the calculation; in the example, this is at −20 dB with respect to the main lobe. Figure 5.8 shows the corresponding spectral distribution of clutter with respect to S_{min} for the values assumed, indicating that even the side-lobe clutter could be troublesome.

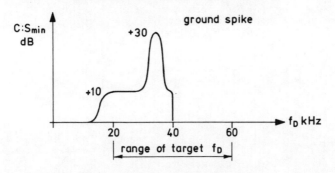

Figure 5.8   *Spectral distribution of clutter*

References

1. Skolnik, M. I., *Introduction to Radar Systems*, 2nd edn, McGraw-Hill, 1980, Section 13.2.
2. Skolnik, M. I., *Introduction to Radar Systems*, 2nd edn, McGraw-Hill, 1980, Chapter 13.
3. Nathanson, F. E., *Radar Design Principles*, McGraw-Hill, 1969, Chapter 7.
4. Farrell, J. A. and Taylor, R. L., 'Doppler radar clutter', Paper VI-2 in Barton, D. K. (ed.), *Radars, Vol. 7, CW and Doppler Radar*, Artech House, 1978.

6

CW-Doppler Guidance Technology

6.1 Introduction

An outline of the principles of the CW-Doppler seeker has been given already in section 2.3, and derivations of the Doppler frequency spectra appear in chapter 5. The purpose of the present chapter is to present a detailed account of the signal processing in the receiver, of the properties of the illuminator and of the effects of external factors on the guidance system.

The functions of the guidance receiver are to amplify signals received by the antenna, to extract the Doppler frequency of every signal, and to select the wanted target echo on the basis of its Doppler frequency, excluding all others; steering signals are then derived from the angle tracking information conveyed by the selected target echo. Outline block diagrams of active and semi-active seekers appear in figures 2.3b and 2.4b respectively; however, several different receiver configurations have evolved during the 30 years or so of the existence of the technology[1,2], and the next four sections of the chapter describe these configurations in a comparative manner. Various terms are employed and it will be helpful to explain these at the outset.

Late narrow-banding	The wanted target echo is segregated in Doppler frequency from all other signals by a narrow band-pass filter at a late stage in the receiver chain incorporated within the main IF amplifier (IFA)
Early narrow-banding (also known as an 'inverse receiver')	The segregation takes place as early as possible in the receiver chain, before the main IF amplifier

Doppler Frequency Extraction

Explicit	The Doppler frequencies are translated to baseband and hence each Doppler frequency is *explicitly* represented by an AC frequency
Implicit	The Doppler frequencies remain always as shifts on a carrier and their presence is *implied* by the change of frequency

Doppler Tracking

Frequency lock loop	The Doppler frequency of the wanted target echo is changed to a second intermediate frequency which is compared with that set by a fixed frequency discriminator. Any maladjustment in frequency is sensed and used to retune the associated local oscillator
Phase lock loop	As for the frequency lock loop, but a phase discriminator is used which senses phase difference. This difference retunes the local oscillator via an integrator

The treatment of the subject is to give first of all a description of Doppler frequency extraction by the late narrow-banding explicit receiver, as this is the earliest type and the techniques are well established. Then follows an account of Doppler frequency selection and tracking; although this is given in connection with the first type of receiver, it relates to all types of receiver configuration. This is followed by the late narrow-banding implicit receiver and then the early narrow-banding receiver, which is implicit by nature. These sections conclude with a block diagram of a complete seeker using early narrow-banding.

6.2 Late narrow-banding receivers – explicit Doppler extraction

The most common arrangement of this is as shown in figure 6.1. The target signals picked up in the front antenna are amplified at intermediate frequency (IF) by the superheterodyne principle in the channel marked 'Signal IFA'. The reference signal for extraction of the Doppler frequency is derived from the original transmitted signal, either directly by means of a low-power transmission line coupler, as in an active system, or indirectly via the rear reference transmission in semi-active homing. The reference signal is converted to the same IF as that of the target signals by a common microwave local oscillator, and it is then amplified in a separate reference IF channel, the 'Ref IFA'.

Doppler extraction is by heterodyning the reference signal at the output of the 'Ref IFA' with the target signals at the output of the 'Signal IFA' in the Doppler balanced mixer; the target signal Doppler frequencies appear at the output of this mixer at baseband. The signal IFA carries out a large part of this amplification of the target signals, with a maximum gain of about 100 dB. AGC

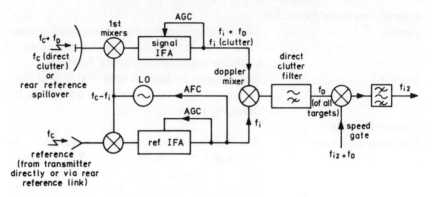

Figure 6.1   *Explicit Doppler receiver (IF stages)*

is applied to compensate for the large increase in target signal strength as the missile closes on the target, being set usually so as to maintain the signal IFA output at about -10 dB or less with respect to that of the reference IFA to ensure optimum conditions for the Doppler mixer.

The Doppler frequency band is typically 5–100 kHz, but the receiver bandwidth is several times this to avoid ill effects of non-linearities in the receiver. This does not degrade the ultimate signal: noise ratio as the noise bandwidth of the receiver is that of the Doppler gate later in the receiver chain. The actual IF has one of the typical radar values, say 45 MHz. For satisfactory operation of the Doppler mixer with semiconductor diodes, the reference signal applied to it should be at least 1 volt RMS. To avoid undesirable forms of cross-modulation, the reference IFA channel should be identical with the signal channel except, of course, for gain. In a fully active system this gain can be fixed, but in a semi-active one AGC must be applied to compensate for the falling-off in reference strength as the missile flies away from the illuminator.

The AFC is applied to the local oscillator by comparing the reference channel output frequency with that of a frequency or phase discriminator tuned to the first IF. The loop has to perform two functions: to act as a conventional AFC to hold the reference signal in the centre of the IFA passband, and to counteract high-frequency, small-deviation, frequency modulation of the local oscillator induced by vibration. Some early systems employ two separate loops, but modern practice is to combine the functions in one.

It is impossible to avoid breakthrough of the transmitted signal in an active system through the duplexer into the front of the signal amplifier chain; equally it is impossible to avoid breakthrough of the rear reference signal in a semi-active system by diffraction over the surface of the missile and into the front antenna. This reference spillover or direct clutter (so called because it has no Doppler shift with respect to the reference signal) is often very large, say +90 dB with respect to the wanted target signal at the beginning of the engagement. In a late narrow-banding system, the direct clutter cannot be removed until after Doppler extraction; it is then removed fairly easily in an explicit receiver by a sharp cut-off high-pass filter after the Doppler mixer.

An ordinary balanced (or zero IF) mixer does not distinguish between positive and negative Doppler shifts — that is, a signal in the Doppler video channel could be from either an approaching target, or from one receding from the missile. Successful engagement of the latter would be impossible, so negative Doppler frequencies are eliminated by a conventional single sideband circuit, as shown in figure 6.2. Rejection of negative Doppler frequencies is about 12 dB in practice; there is also an improvement of nearly 3 dB in receiver noise figure, owing to elimination of the noise from the lower side-band channel.

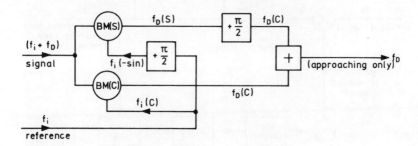

Figure 6.2   *Non-ambiguous second mixer*

6.3 Doppler selection and tracking

Introduction

The output of the Doppler mixer described in section 6.2, after the clutter filter, contains the Doppler frequencies of all targets in the beam of the front antenna. Of these, only one is wanted for the engagement and it is picked out from all the others by its unique Doppler frequency. This selection is carried out by a narrow bandpass Doppler filter or gate.

To cope with inevitable changes in closing velocities and hence in Doppler frequency, the gate is controlled by an AFC circuit so that it tracks the Doppler frequency of the selected target. To avoid having to tune the narrow-band gating filter, the superheterodyne principle is again employed and the wanted Doppler frequency is translated to the fixed second IF of the gate, usually 100 kHz or so, by mixing with a tunable local oscillator controlled by the AFC circuit.

Automatic frequency control is either by frequency discriminator (frequency lock loop) or by phase discriminator (phase lock loop) and these will be described separately.

Frequency lock loop

The block diagram of a typical loop is shown in figure 6.3. Under static conditions — that is, f_D constant — the loop is perfectly balanced and the difference between the local oscillator frequency, f_0, and the Doppler frequency of the wanted target is exactly f_{i_2}, the design frequency of the narrow bandpass filter gate. All other targets of different Doppler frequency are excluded; only the wanted target echo is allowed through to influence the steering of the homing head and of the missile.

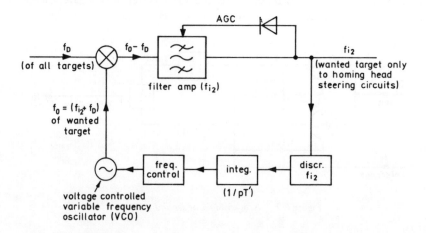

Figure 6.3   *Frequency lock loop — loop balanced*

If the gated target's Doppler frequency changes by a small amount, $(f_0 - f_D)$ no longer equals f_{i_2} and there is an imbalance in the loop. This is sensed by the frequency discriminator which generates an error voltage which is used, via an integrator, to change f_0 to restore balance. This situation is depicted in figure 6.4 which shows the voltages or frequencies present at each point in the loop.

The following simple mathematical analysis shows that the loop has a control differential equation of the first order and that a constant rate of change of Doppler frequency (that is, a constant closing acceleration of target and missile) leads to a frequency tracking error.

The loop equation is

$$f_0 = f_{i_2} + \frac{K_1 K_2 \, (f_{i_2} - f_0 + f_D)}{sT^1} \; ; s \equiv \frac{\mathrm{d}}{\mathrm{d}t} \tag{6.1}$$

Putting $\dfrac{T^1}{K_1 K_2} = T$ and re-arranging terms gives

$$f_0(1 + sT) = f_{i_2} + f_D$$

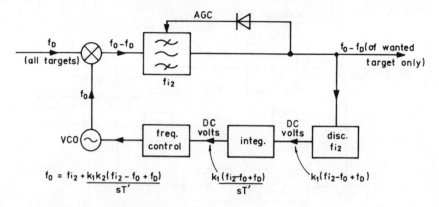

Figure 6.4   *Frequency lock loop – loop under tracking conditions*

Hence

$$f_0 - f_D = f_{i_2} - \frac{sTf_D}{1 + sT} \tag{6.2}$$

If the target has constant closing velocity, f_D is constant and $(f_0 - f_D) = f_{i_2}$, as in figure 6.3. If the target has radial acceleration then there is a tracking error in frequency of Tf_D. For example, acceleration = 50 m s^{-2}, λ = 22 mm, f_D = 4.5 kHz s^{-1}, and T = 0.1 s, giving a frequency error of 450 Hz.

The filter bandwidth is the effective noise bandwidth of the receiver and this should therefore be as narrow as possible; it should also be narrow to exclude adjacent targets. On the other hand, it must be wide enough to accommodate the wanted target frequency spectrum (including conical scan sidebands, if present), to accommodate acceleration lags and for rapid initial acquisition. In practice the width is a few hundred hertz.

The loop time constant T should be short enough to ensure speedy response to a changing Doppler frequency, but long enough to keep jitter down; practical values are 0.01–0.1 s.

Phase lock loop[3]

In this, not only is the wanted target signal converted to the second IF (f_{i_2}) – that is, locked in frequency – but it is also locked in phase. This eliminates frequency error in the presence of closing acceleration of missile and target, and thus the velocity gate width may be narrower; a few tens of hertz are practicable. The advantages of this compared with the frequency lock loop are better receiver sensitivity (narrower noise bandwidth) and better discrimination between targets of adjacent Doppler frequency; the disadvantage is slower acquisition. Figure 6.5 shows the block diagram of the loop in the static condition – that is, at constant

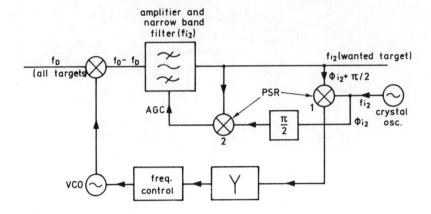

Figure 6.5   *Phase lock loop – loop balanced*

f_D. The loop locks in phase with a $\pi/2$ difference between the phase of the gated target and that of the local crystal oscillator. AGC voltage is derived from the gated target signal which is converted to DC in a second phase-sensitive rectifier or detector switched from the crystal oscillator. The $\pi/2$ network cancels out the phase difference referred to above, to bring the two inputs into phase.

The receiver noise bandwidth is still that of the Doppler frequency narrow band pass filter but, because there is now no need to accommodate dynamic lags in frequency, this bandwidth can be much narrower, 100 Hz or even less. This feature gives much better resolution in Doppler frequency and it is even possible to discriminate between two targets flying close together in formation.

The following mathematical analysis shows that the loop has a second order transfer function and that a constant rate of change of Doppler frequency leads to a phase error but to no frequency error.

Figure 6.6 shows the loop under dynamic tracking conditions. Because of a changing f_D, there is now a phase error; the phase difference is no longer necessarily $\pi/2$. At PSR 1

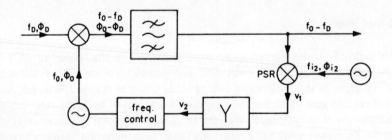

Figure 6.6   *Phase lock loop – loop under tracking conditions*

$$V_1 = K_1 \cos(\phi_{i_2} - \overline{\phi_0 - \phi_D}) \text{ where } K_1 \text{ is a constant}$$
$$= K_1 \sin(\phi_{i_2} - \overline{\phi_0 - \phi_D} + \pi/2)$$

This is depicted in figure 6.7 as a graph.

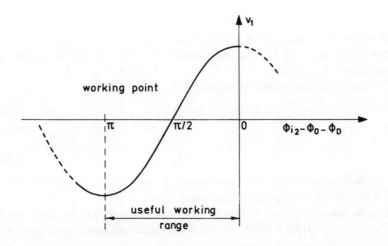

Figure 6.7   *Phase lock loop – graph of V_1 as a function of ϕ_D*

The working range is that over which the graph has a positive slope over $\pm\pi/2$ of phase error. In practice, phase errors are usually well within these limits and the sine can be taken to be equal to the angle, permitting linear analysis.

$$V_1 \approx K_1 (\phi_{i_2} - \overline{\phi_0 - \phi_D} + \pi/2)$$

From figure 6.6

$$V_2 = YV_1, \text{ where } Y \text{ is a network transfer function}$$

and $(f_0 - f_{i_2}) = K_2 V_2$, where K_2 is another constant.
  By the action of the frequency controller

$$f_0 - f_{i_2} = \frac{s}{2\pi} (\phi_0 - \phi_{i_2} + A)$$

where A is some phase constant, and $s \equiv \mathrm{d}/\mathrm{d}t$. Combining this with the first two equations gives the loop control equation

$$\frac{s}{2\pi} (\phi_0 - \phi_{i_2} + A) = K_1 K_2 \, Y \left[\phi_{i_2} - \overline{\phi_0 - \phi_D} + \frac{\pi}{2} \right] \qquad (6.3)$$

Put $2\pi K_1 K_2 = K$; choose Y to be of the form $\left(1 + \dfrac{1}{sT} \right)$, which can be realised by the network of the form shown in figure 6.8, where $T = CR$. Whence the control equation becomes

$$(\phi_0 - \phi_{i_2})(Ts^2 + KTs + K) = K(Ts + 1)(\phi_D + \pi/2)$$

This is of the form

$$(\phi_0 - \phi_{i_2})(s^2/\omega_n^2 + 2\zeta/\omega_n + 1) = (2\zeta/\omega_n + 1)(\phi_D + \pi/2) \qquad (6.4)$$

where $\omega_n^2 = K/T$, and $\omega_n/2\zeta = 1/T$, in which ω_n is the undamped natural frequency of the loop and T is the time constant; ζ is the damping coefficient. If f_D is constant, $\phi_D = 2\pi f_D t$.

The steady state solution of equation 6.4 is

$$\phi_0 - \phi_{i_2} = 2\pi f_D t + \pi/2 \qquad (6.5a)$$

and

$$f_0 - f_{i_2} = f_D \qquad (6.5b)$$

that is, the loop locks with no frequency error and with the fixed phase difference of $\pi/2$, which corresponds to zero phase error.

If f_D increases uniformly with time so that $f_D = (f_{D_0} + \dot{f}_D t)$, the corresponding phase, ϕ_D, is $2\pi(f_{D_0} t + \dot{f}_D t^2/2)$. The steady state solution of equation 6.4 is

$$\phi - \phi_{i_2} = 2\pi[f_{D_0} t + \dot{f}_D t^2/2 + \pi/2 - \dot{f}_D/\omega_n^2] \qquad (6.6)$$

There is still no frequency error but there is a phase error of $-2\pi\dot{f}_D/\omega_n^2$, which must not exceed $\pi/2$, as figure 6.7 shows. For example, $\dot{f}_D = 2$ kHz s^{-1} requires that $\omega_n \geqslant 90$ rad s^{-1}. ζ controls only the damping of the transient response and will lie probably between 0.5 and 0.7 (critical damping). Further information on the theory of linear closed loop systems can be found in most textbooks of servomechanisms.

If, as is usual during tracking, there is no error in the frequency setting of the voltage-controlled oscillator VCO, the output of PSR 1 can only be DC, corresponding to a phase error. A glance at the circuit diagram of Y shows that its gain to DC is very high, so that the full effect of the error is felt by the frequency controller. This ensures that the frequency generated by the VCO remains correct, even in the presence of a changing f_D. The closing acceleration is, however, almost certain to change at some time during an engagement, even if only momentarily. A change of phase error constitutes a frequency error, so the loop must have some response to frequency errors. This response is determined by the frequency response of Y, which is of the form shown in figure 6.8b.

The useful response of the loop is confined to frequency errors of certainly less than $1/2\pi T$; if $T = 0.01$ s, this is less than about 15 Hz. This is admirable for tracking but would make the process of initial target acquisition by the loop very tedious. Acquisition is usually by the process described in more detail below – breaking the loop temporarily and causing the VCO to sweep about a predicted value, until f_D of the wanted target appears in the Doppler frequency

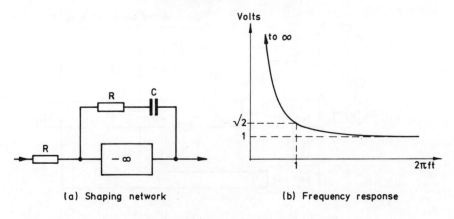

(a) Shaping network   (b) Frequency response

Figure 6.8   *Phase lock loop*

gate. The width of the gate being much more than the bandwidth of the loop, the sweeping process still has to continue until $(f_0 - f_D)$ is nearly equal to f_{i_2}. The response of Y is then sufficiently large for the loop to be able to take over and to remove the remaining frequency error. The sweep rate must not be so fast that the VCO is swept over the correct value before the loop has time to appreciate the presence of a target and to lock on to it. In view of the very narrow band of frequency errors over which acquisition is possible, the sweep rate must be slow enough for each of these increments of frequency to be examined in turn. Bearing in mind that the total ambit of search of the VCO may be several kHz, searching and acquisition would be far too long a process. The difficulty can be alleviated either by using a much smaller value of T for acquisition, thus widening the band of frequency errors to which the loop will respond, or acquiring the target initially by a subsidiary frequency lock loop which then transfers it to the phase lock loop with only a small residual frequency error. For further information on phase lock loops, see ref. 4.

Automatic acquisition of the Doppler frequency of the wanted target (see figure 6.9)

As explained later in section 6.10, Doppler tracking systems cannot hold lock readily through boost, because of high Doppler rate of change, vibration of the whole missile and, in semi-active homing, because of excessive flame modulation of the rear reference; hence it is necessary to re-acquire the target after the conclusion of the boost phase. This paragraph describes the acquisition circuit.

The Doppler frequency tracking loop is set before launch to the value of f_D predicted to occur at the start of acquisition. The process of acquisition consists of the Self-run circuit sweeping the VCO for f_D about the pre-set value until a signal passes through the narrow band filter. If this signal is strong enough to

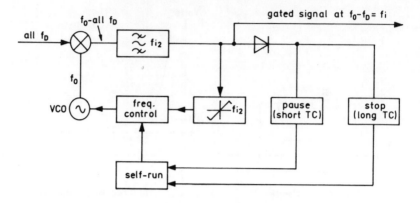

Figure 6.9   *Automatic acquisition circuit*

exceed a threshold, the Pause circuit stops the sweeping for a few tens of milli-seconds so that the Stop circuit can examine the signal. If it persists, the signal is taken to be a genuine target echo and not just an exceptionally high noise peak, and the Stop circuit inhibits the Self-run permanently. The frequency tracking loop then takes over. Should the target fade subsequently, the Stop circuit still prevents Self-run for a second or so to give the target signal a chance to re-appear. Should it not, the Stop circuit releases the Self-run, which starts re-searching about the last known value of f_D.

In some systems, mere persistence is not taken as sufficient evidence that the correct target has been acquired. The Stop circuit can be made to look for a modulation on the echo of the correct frequency and phase. The presence of this confirms the presence of the correct target. This is the procedure known as 'coherency' or 'confirmation check'.

Acquisition could be speeded up by using a bank of filters covering the whole search ambit simultaneously; this is usually too cumbersome in analogue signal processing, but is attractive with digital signal processing using the Fast Fourier Transform as described in chapter 7.

6.4 Implicit Doppler extraction (see figure 6.10)

In an explicit receiver the target signal is translated to baseband before the narrow band filter. Within the baseband there may be many spurious signals generated within the missile, such as harmonics of the power supply frequency and vibration induced EMF. It is difficult to prevent the baseband amplifier from picking these up and presenting them to the Doppler acquisition and track-ing circuits as genuine target signals. In the implicit receiver there is no electrical signal at the Doppler frequency. The Doppler frequency is 'implied' in the change of some carrier frequency, and the wanted target signal at $(f_c + f_D)$ is

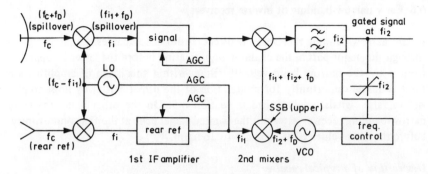

Figure 6.10 *Implicit Doppler receiver*

translated firstly to $(f_{i_1} + f_D)$ and subsequently to f_{i_2}, the frequency of the narrow band filter. By avoiding amplification at such low frequencies, the implicit receiver is not faced with this difficulty; on the other hand, the explicit receiver has the advantage that it can subject the direct clutter to an extra stage of filtering, in the clutter filter. The implicit receiver is, of course, non-ambiguous.

6.5 Generation of false target signals by spurious noise modulation

In late narrow-banding receivers, such as that of figure 6.1, the direct clutter (rear reference spillover or transmitter breakthrough) and indirect clutter (ground reflection) accompany the wanted target signal in the chain of amplification. This poses difficulties, though not insuperable problems, of dynamic range and of AGC within the receiver.

A more intractable difficulty is the creation of false target signals of finite Doppler frequency within the receiver chain by the generation of sidebands by vibration of the local oscillator. These sidebands are transferred on to the direct or indirect clutter at the first mixer in the signal channel of the receiver. Each one of them is readily interpreted by the Doppler mixer as a Doppler frequency; if the vibration frequency lies within the operational Doppler frequency band, and if the strength of the spurious signal is great enough, there is a risk of it being mistaken for a genuine target signal by the Doppler acquisition and tracking circuit. Both amplitude and angle modulated sidebands are responsible but angle-modulated sidebands are the more troublesome and their suppression gives rise to the need for the fast-acting AFC loop referred to in section 6.2. The same effect is produced by noise sidebands on the illuminator transmission, so that noise suppression circuits are usually needed here as well[4]. A fuller description of the effects of angle-modulated noise is given in appendix B.

6.6 Early narrow-banding or inverse receivers

In late narrow-banding receivers the direct clutter accompanies the target signals through the major part of the chain of amplification before the latter is separated from it by the narrow band velocity filter. Within this chain of amplification there is ample opportunity for modulation of the direct clutter by noise to give rise to false signals, as described in section 6.5. On the other hand, the early narrow-banding receiver gates-out the wanted target echo at the first opportunity, before the direct clutter has a chance to do harm.

Description of a typical receiver

Figure 6.11 shows a possible form of this type of receiver. It will be seen straight away that the narrow band Doppler filter is at the head of the signal IF amplifier chain. In this position it removes all unwanted signals, including the direct clutter, before they can do any harm by cross-modulation or overloading. Until recently, the difficulty with this system has been the manufacture of a narrow bandwidth filter with a steep rate of cut-off at such a high frequency as the first IF, but this difficulty has now been overcome and it is in use in modern missiles.

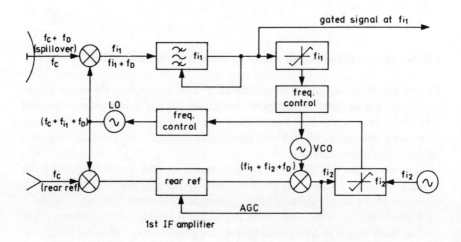

Figure 6.11 *Early narrow-banding or inverse receiver – balanced condition*

Detailed account of the working of the circuit

There are two frequency tracking circuits working together – one in the front channel and one in the rear; unlike as in the late narrow-banding receivers, the Doppler frequency shift due to the wanted target is transferred to the rear reference signal. The wanted target echo in the front receiver is translated to the fixed frequency of the first IF, placing it within the pass band of the narrow band Doppler filter. Under the static conditions shown in figure 6.11, where there is no closing acceleration, f_D is constant and there are no dynamic lags in frequency or in phase. Figure 6.12 shows the situation when there is closing acceleration and f_D is changing.

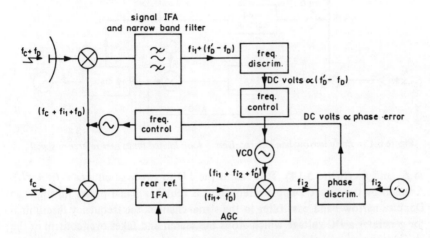

Figure 6.12   *Early narrow-banding receivers − tracking condition*

The time constant of the front (target) loop is of the order of milliseconds, just short enough to deal with missile–target accelerations; that of the rear (rear reference) loop is as short as a few microseconds to ensure rapid acquisition of the rear reference signal. Although there are two time constants in the system, the vast difference in their magnitude reduces the system transfer function to the first order.

A frequency lock loop is chosen for the front loop on the grounds of ease of manufacture, though at the cost of a slightly inferior performance to that obtainable with a phase lock loop. If desired, a phase lock loop with an extra-narrow band pass filter can be added in tandem to improve the Doppler frequency resolution − for example, for multiple target discrimination.

Initial acquisition of the target

The first task of the receiver is to acquire the rear reference. This can be done by setting the rear reference to the fixed frequency $(f_{i_1} + f_{i_2})$ and then sweeping the LO about the estimated value of $(f_c + f_{i_1})$ (figure 6.13). As soon as the correct value of f_{i_1} appears in the rear receiver, the phase discriminator produces a DC voltage which stops the sweep of the LO and, via the frequency control, makes the final adjustment to the generated $(f_c + f_{i_1})$ to bring f_{i_1} into the correct phase. The LO is now locked to the rear reference.

The VCO controlling the frequency of the rear reference local oscillator now takes over and sweeps it about the predicted value of Doppler frequency, f_D' added

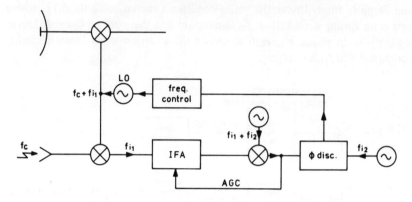

Figure 6.13   *Early narrow-banding receivers – loop locked on to rear reference signal*

to f_{i_1} and f_{i_2} (figure 6.14). This causes the LO to sweep about $(f_c + f_{i_1} + f_D')$. As soon as the target Doppler frequency is found, a signal passes through the Doppler narrow band pass filter in the signal channel; the frequency discriminator generates a DC voltage which stops the search and takes over control of the VCO by the associated frequency controller.

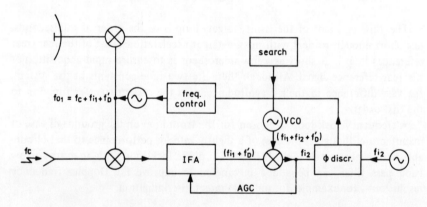

Figure 6.14   *Early narrow-banding receivers – searching in Doppler*

6.7 Complete guidance receiver

Figure 6.15 is the block diagram of a complete two-plane inverse monopulse receiver with three separate channels, showing the interconnection of the various functions described separately already. In practice, there would be two or more frequency changes in the IF amplifiers to avoid instability.

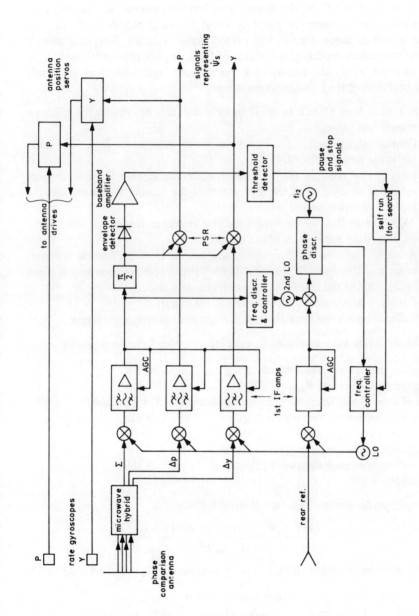

Figure 6.15 *Inverse monopulse seeker – block diagram*

6.8 Rear reference and the illuminator

The main purpose of the illuminator is self-evident; a secondary purpose is to provide a coherent reference signal of about 1 volt amplitude at the Doppler mixer of the guidance receiver. In an active homer a simple directional coupler on the transmitter feeder, plus a means of fixed amplification at IF, suffices. In a semi-active homer there must be a rear reference transmission to the missile in flight, fulfilling the following requirements

(a) *S:N* of at least +10 dB at the Doppler mixer; this determines the reference transmitted power.
(b) Constant signal level at the Doppler mixer throughout the flight; this determines the range of the AGC.
(c) Field of view from the rear reference transmitting antenna at the illuminator sufficient to encompass the movements of the missile; this determines the antenna beamwidths and gain.
(d) An adequate field of view from the rear reference antenna on the missile, encompassing manoeuvres of the missile.
(e) A method of avoiding destructive interference between the rear reference beam and the main illumination beam or its side-lobes. One way is phase modulation of the rear reference beam at a slow rate.
(f) Circular polarization to allow for roll of the missile.
(g) Sufficient power to work through the missile sustainer motor flame.

The following is an example of a calculation arising from these requirements:

Maximum range, illuminator–missile, R_{max}	50 km
Minimum lock-on range, R_{min}	2 km
Field of view of rear reference transmitting antenna	0.4 × 0.8 rad
Missile rear reference antenna gain, G_m	3 dB
Rear reference receiver noise figure, NF	10 dB
Miscellaneous losses, including flame attenuation, L	2 dB
Rear reference receiver bandwidth, B_r	2 MHz
Wavelength, λ	22 mm

The equation for the rear reference transmitted power is

$$P = \frac{(4\pi)^2 \, L \, R_{max}^2 \, NF \, (k \, T_0 \, B_r) \, (S/N)_{min}}{G_I \, G_m \, \lambda^2}$$

Illuminator antenna gain is given by the approximation; $G_I = 4\pi/(0.4 \times 0.8) = 39$

$$(S:N)_{min} = + 10 \text{ dB}$$

$$P = \frac{(4\pi)^2 \times (5 \times 10^4)^2 \times 16 \times (4 \times 10^{-15} \times 2) \times 10}{39 \times 2 \times (22 \times 10^{-3})^2}$$

$$= 13.4 \text{ W}$$

This probably represents about 5 per cent of the total illuminator power radiated.

The receiver AGC must maintain a constant output signal between 2 and 50 km; since signal strength is inversely proportional to R^2, the AGC range to cater for this = $20 \log_{10}(50/2)$ = 28 dB. In AAM semi-active homers, there is usually no room on the parent aircraft for a separate rear reference antenna and the missile has to rely on the side-lobes of the illumination. These may be cross-polarized with respect to the main beam, but if the missile rear reference antenna is circularly polarized this presents no special difficulty.

Illuminator noise

Noise on the rear reference transmission is equivalent to noise on the local oscillator, as mentioned in section 6.5. The noise performance of the transmitter has to be monitored carefully[5]. The permissible limit of FM noise is the same as for a local oscillator — that is, for the example given in appendix B, the limit would be 320 Hz deviation anywhere within the sideband of expected Doppler frequencies. Similarly, for AM noise, if the ratio direct clutter:S_{min} were + 90 dB, no AM noise within the Doppler filter bandwidth should exceed − 90 dB with respect to the carrier. The rear reference is also a useful communication link to the missile in flight, conveying information and commands in the form of modulations on a sub-carrier. Such commands are 'arm the warhead', 'self-destroy', 'change navigation constant' etc. In Target via Missile (TVM) the rear link is two-way. The down link conveys guidance information from the guidance receiver; the up link conveys steering commands to the control system. The signal processing takes place at the parent site, permitting the use of a much more elaborate signal than the guidance receiver could interpret on its own, and thus increasing the resistance to ECM or to mutual interference.

The rear reference may be omitted altogether provided that the transmitter frequency and the local oscillator frequency are sufficiently stable for the velocity gate to be able to hold lock. For example, the maximum permissible rate of change of frequency for a phase lock loop is $\omega_n^2/4$ (section 6.3); if ω_n = 90 rad s^{-1} the rate is 2 kHz s^{-1}, and at a carrier frequency of 16 GHz this is 1 part per second in about 10^8. Such a stability is obtainable now with solid state quartz crystal-controlled sources driving a chain of frequency multipliers.

Miscellaneous aspects of the illuminator

In some SAM systems and in all AAM systems the illuminator also acts as target tracker. This does not affect its role as an illuminator as this is given first consideration; however, CW transmission is not very convenient for a target tracker.

In a SAM system the tracker/illuminator would need separate antennas for transmission and reception, well screened from each other. In AAM, where there is insufficient room for two antennas, an elaborate method of duplexing, known as 'feed-through nulling'[6] might have to be employed. Modern practice is either to employ a separate pulse transmission at a different frequency, sharing the single antenna with the CW illumination, or to use pulse-Doppler.

The illumination is often circularly polarized so that the signal received by the seeker is unaffected by roll of the missile. If the missile is roll-stabilized, vertical polarization is used so as to reduce the image effect with low-flying targets. The illumination is sometimes coded — for example, by FM; the guidance receiver looks for correlation between the code on the target echo and that on the rear reference, thus enabling the seeker to recognize signals from its own illuminator (coherency check).

6.9 Rocket motor flame effects on microwave transmissions

Microwave transmissions to the rear of a missile usually have to pass through at least part of the length of the motor flame. In homing missiles this concerns mainly the rear reference link in semi-active homers, though it would also concern any command link. The hot flame of the motor consists of a plasma of ionized particles, especially free electrons; it is these which are mainly responsible for interference with microwave communications, by attenuation of the signal and by the introduction of noise-like amplitude and phase modulations.

This section describes the effects of the rocket motor flame on microwave transmissions to the rear of a homing missile, and the subsequent effects on the performance of the guidance receiver.

When an electro-magnetic wave passes through a plasma of free electrons, it interacts with the electronic charges to set the electrons in motion. Some of the kinetic energy thus acquired by the electron is lost by collisions and it is this effect which is responsible for the attenuation. The storage of the remaining energy gives rise to reactive effects in the plasma medium which introduce phase changes into the wave. However, these effects are not constant, because of the highly turbulent flame; this turbulence imposes random amplitude and phase modulation on to the wave passing through. The effects are measured by a suitable microwave test rig; typical values for the attenuation of a pair of small motors appear in table 6.1; associated modulation noise spectra are shown graphically in figure 6.16. Since the amplitude and phase noise spectra are usually identical, or nearly so, figure 6.16 applies to both; note the beneficial effect of suppressing 'flash' or 'after burning'.

Table 6.1. Attenuation (at 10 GHz) of typical small solid motors

Sustainer, flash suppressed	1 dB
Sustainer not flash suppressed	5–10 dB
Boost, not flash suppressed	15 dB or more

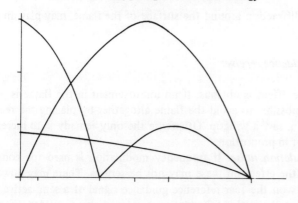

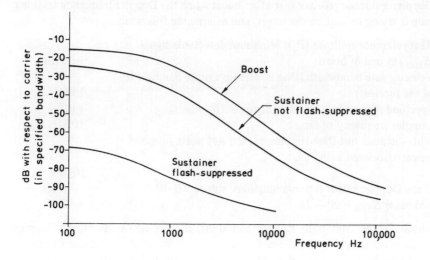

Figure 6.16 *Motor flame noise spectra*

Effect of frequency

Theoretical studies of plasma indicate that attenuation should stay constant with increasing frequency and then start to fall[7]; in practice, it increases as is shown by the following typical figures for the difference in attenuation

$$16\,\text{GHz} : 10\,\text{GHz} \quad +5\,\text{dB}$$
$$35\,\text{GHz} : 16\,\text{GHz} \quad +5\,\text{dB}$$

Similar differences occur in the noise spectra.

The contradiction is attributed to the fact that a rocket motor flame of limited dimensions is not the infinite uniform plasma assumed in the theory. Other effects, such as diffraction around the surface of the flame, may play an important part.

Effects on the guidance receiver

Attenuation. The effect is obvious. If no improvement in the flame is possible then it may be possible to avoid the flame altogether by placing the rearward-facing antenna on, say, a wingtip. Otherwise the only remedy is to increase the transmitter power in proportion.

Amplitude modulation noise. If frequency modulation is used for commands over a rear link, the effect on these may not be serious. There may however be significant effects on the rear reference guidance signal of a semi-active CW or pulse-Doppler homer, as the following examples show. Consider an explicit Doppler guidance receiver just after boost when the Doppler frequency tracking loop is trying to acquire the target, and assume the following

Rear reference spillover (P_c): Minimum detectable signal (S_{min}) at end of boost	$+90$ dB
Velocity gate bandwidth (that is, effective noise bandwidth of the receiver)	1 kHz
Specified narrow bandwidth of figure 6.16	1 kHz
Doppler frequency of target	10 kHz
With sustainer not flash-suppressed, the AM noise sideband power associated with P_c is	-75 dB (figure 6.16)
If the Doppler mixer is non-ambiguous, subtract 3 dB	-78 dB
AM noise: $S_{min} = 90 - 78$	$+12$ dB

Above 30 kHz the ratio is less than 0 dB, so that echoes at higher Doppler frequencies should be unaffected. On the other hand, the noise can be modulated also on to the ground spike (section 5.2); since this is not at zero Doppler frequency, the noise spectrum is translated to a higher band where it can now interfere with higher Doppler frequency target signals. Note that flash suppression eliminates the difficulty with the rocket motor quoted, at least for likely Doppler frequencies for airborne targets.

Frequency modulation noise. Section B.3 of appendix B quotes the AM noise sideband RMS amplitude due to one cause of conversion of angle modulated noise as $V_c \tau \Delta \omega$; expressed as noise power in a bandwidth B this becomes $P_c (\tau 2\pi \Delta f)^2$, where P_c is the rear reference spillover power (or ground spike power), τ is the differential time delay between signal and reference IF channels and Δf^2 is the mean square noise frequency deviation in the bandwidth B. Provided that the frequency to which the filter of bandwidth B is tuned is much less than f_D, the ratio of phase modulated noise sideband in B:carrier power $= \frac{1}{2}(\Delta f/f_D)^2$; using the numerical data given above, at $f_D = 10$ kHz, sustainer not flash suppressed

$$\Delta f^2 = 10^{-7.5} \times 2 \times 10^8$$
$$= 2 \times 10^{\frac{1}{2}} \text{ Hz}^2$$

Assuming $t = 0.5$ μs, the converted AM noise power: minimum detectable signal is

$$10^9 \times 2 \times 10^{\frac{1}{2}} \times 4\pi^2 \times (0.5 \times 10^6)^2$$
$$= -12 \text{ dB}$$

6.10 Initial acquisition of the target

The seeker must acquire the target signal and lock on to it so that it can track in the coordinates of interest, which in CW-Doppler are the two angles and the velocity. In 'all-the-way' homing the procedure would appear to be straight-forward; the parent radar will be tracking the target and will be able to pass the target coordinates directly to the seeker. Unfortunately, it is difficult for the guidance receiver to hold Doppler frequency lock through the boost phase; the first time-derivative of this frequency is very large because of the large acceleration, many 10s of g, creating a large dynamic lag in the frequency track-ing servo system. It is possible to allow for this by using an especially wide velocity gate during boost, but this increases the receiver noise just at the time when the signal is at its weakest. It is also possible to programme the velocity gate from the missile accelerometer to compensate for the acceleration; this is satisfactory provided that the boost dispersion is not too great, otherwise missile acceleration may no longer represent adequately the closing acceleration. A second difficulty is the rise in noise level resulting from the strong missile vibra-tions; not only is there a general rise, degrading the $S{:}N$, but there are likely to be strong spectral lines corresponding to particularly strong vibrations. The velocity loop could easily transfer lock to one of these. Semi-active homers have to contend also with the effects of passage of the rear reference signal through the violent flame of the boost motor. This creates attenuation, general noise, and individual noise spectral lines. For these reasons, lock-on is delayed usually until the end of the boost phase. Before launch, the seeker antenna is pointed in the direction in which it is predicted that the target will lie at the end of boost, and it is stabilized in this direction by the antenna rate gyroscope. Because the boost phase is short, only a few seconds, and because the corresponding distance travelled by the missile is short, the target should still be in the field of view of the antenna, thus obviating the necessity for a search in angle. For example, consider a missile which accelerates from 0 to 800 ms^{-1} in 2 s and which is launched against a target at a range of 20 km. The distance travelled during boost is 800 m; even if the boost dispersion were as much as 30°, the angular dispersion of the sight line would be only 20 mrad, which should be well within the antenna half-power beamwidth. Another possible cause of angular dispersion of the sight line is drift of the gyroscope; a typical drift rate is 1 rad h^{-1}, which is, however, negligible over a few seconds.

Similarly, the Doppler frequency is set before firing to the predicted value at the end of boost. Unfortunately, it is not possible to predict to within a gate width, and searching is necessary. Consider the same missile and assume a wavelength of 22 mm and a gate width of 500 Hz, corresponding to a velocity increment of 11 m s^{-1}. Assume that a variation in boost acceleration leads to an error of 80 m s^{-1}; an angular dispersion of 30° leads to an additional error of 107 m s^{-1}. Assuming that these are extreme variations and that the two are uncorrelated, it would be necessary to search ± 134 m s^{-1} — that is, ± 12 gate widths about the predicted value. The rise time in a filter of width 500 Hz is about 2 ms; allowing a factor of 10 for integration gives a dwell time of 20 ms on each 500 Hz increment, giving a total search time of 0.48 s. In this time the missile would travel a further 400 m, which is not of consequence. Hence the accuracy of predicting the Doppler frequency is adequate.

Vertical launch

During the boost phase the missile rises vertically, turning over gradually into a conventional trajectory. The only additional inconvenience is that it is not possible to orient the antenna before launch. It is therefore locked and the predicted sight line is held on a separate gyroscope until the end of boost. The antenna can then be oriented.

Acquisition in the terminal phase

This presents more difficulty because of the need to predict the direction and the relative velocity of the target and missile at the end of a long mid-course phase. A complete study taking into account all possible conditions requires the aid of simulation and computation, but the following simple treatment will give a feel for the problem. Figure 6.17 shows the geometry in one plane at the commencement of acquisition; the missile M is flying a constant bearing course towards interception of the target T at the point I, hence it is also a diagram of velocities. Mid-course guidance is by a radar R which exercises the dual function of tracking both M and T in angle, range, and range rate. At the end of the mid-course phase, the parent station computes the direction and the relative velocity of T with respect to M from radar data and passes the information to the missile. The error in each of the coordinates determines the search ambit. Consider first the direction θ; the radar measures the position of M and T independently; at the postulated range of 40 km the angle measurement error is likely to predominate, so the error in range measurement will be neglected. From the geometry

$$\sin \theta = (RT/MT) \sin \alpha$$

$$\delta\theta = (RT/MT) (\cos \alpha/\cos \theta) \, \delta\alpha$$

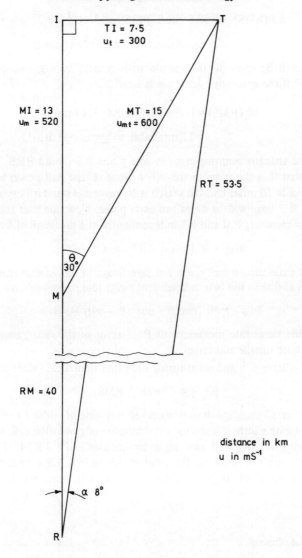

Figure 6.17 *Terminal homing acquisition geometry*

The difference between two independent angle measurements gives the angle α, hence

$$\delta\alpha\ (RMS) = 2^{\frac{1}{2}} \times (RMS\ angle\ error\ of\ the\ radar)$$

Taking the latter as 5 mrad and substituting other numerical data from figure 6.17 gives

$$\delta\theta \text{ (RMS)} = [(53.5 \cos 8°)/(15 \cos 3°)] \times 5 \times 2^{\frac{1}{2}}$$

$$= 29 \text{ mrad due to radar measurements}$$

The angle θ will be given to the missile with respect to a gyroscope axis set before launch; if the gyroscope drift rate is 1 rad h^{-1}

$$\delta\theta \text{ (RMS)} = (1/3600)(40\,000/520) \text{ rad}$$

$$= 21 \text{ mrad due to gyroscope drift}$$

The combined antenna pointing error in this plane is 36 mrad RMS, and it can be assumed that it is the same in the other plane. If the half-power beamwidth of the antenna is 70 mrad and the search is to cover ± 3 standard deviations, the angle search is 3 beamwidths overall in each plane. Now consider relative velocity; the radar measures $\dot{R}M$ and $\dot{R}T$ independently at a precision of δu RMS.

$$u_{MT} = \dot{R}M \cos\theta + \dot{R}T \cos(\theta - \alpha) \tag{6.7}$$

The errors in measuring θ and α are not significant in this case so that the error may be taken as due to the two independent range rate measurements

$$\delta u_{MT} = \delta u \, [\cos^2\theta + \cos^2(\theta - \alpha)]^{\frac{1}{2}} \text{ RMS} \tag{6.8}$$

where δu is the range rate measurement RMS error of the radar, assumed to be the same for both missile and target.

Taking $\delta u = 20$ m s^{-1} and substituting the other numerical values gives

$$\delta u_{MT} = 25 \text{ m s}^{-1} \text{ RMS}$$

In order to cover ±3 standard deviations, a velocity gate of width 11 m s^{-1} must search over 14 gate widths. Hence the total number of resolution cells in a search pattern covering velocity and two angle coordinates is $3 \times 3 \times 14$. If the dwell time per resolution cell is 20 ms, the total search time is 2.5 s. In this time the missile could close on the target by 1.8 km from 15 km, which would be acceptable.

6.11 Terminal phase

In theory a homer must intercept its target but, in practice, random and systematic errors lead to a finite miss distance. This is not necessarily fatal to a mission, as the missile may still strike a vulnerable part of a target of finite size, or the warhead may detonate under the influence of a proximity fuze sufficiently close to be lethal. Nevertheless, it is necessary to make an estimate of errors in order to ascertain the likely effectiveness of the missile. Many of the sources of error lie in minor imperfections in design and manufacture, but the purpose of this section is to examine those which arise from the target and its behaviour as experience has shown that these tend to predominate in the terminal phase. The main sources are glint, multiple targets, and target image; this section treats each of these in turn.

Glint[5]

This is the random wander of the radar centre of a target about a mean position arising from mechanical vibrations, which are inseparable from a moving target. The vibrations cause random changes in the phase relationships between the various scattering surfaces so that the phase centre of the composite echo moves both laterally and longitudinally; since homing is concerned with angles, only the lateral glint is of importance. For a target composed of many scatterers, the distribution of the instantaneous radar centre about the mean is Gaussian with standard deviation usually between one-third and one-sixth of the physical dimension of the target as seen by the radar. Glint is partciularly troublesome to a tracking radar, such as a seeker because, being a low-frequency effect, its spectrum falls within the servo pass band; furthermore, the angle subtended by the glint increases as the missile closes on the target until the glint saturates the missile servo system, and effective guidance and control cease. Fortunately, this occurs only within the last few hundred metres of flight, but nevertheless it can still give rise to a significant component of the miss distance. To evaluate the miss distance due to glint requires a simulator or a computer because of the non-linearity of the missile response because of the limit on the lateral acceleration. However, by making some simplifying assumptions it is possible to gain some numerical insight. Assume that the glint is Gaussian and of standard deviation σ_g metres. In terms of angle at the seeker this is σ_g/R rad, where R is range, in metres. Assuming that the spectral density is uniform between 0 and f_g Hz, the spectral density is $(1/2\,f_g)\,(\sigma_g/R)^2$ rad^2 Hz^{-1}. The spectral density of the rate of change of sight line due to glint is, therefore, $(2\pi f \sigma_g/R)^2/2f_g$ (rad s^{-1})^2 Hz^{-1}, where f is frequency. Assuming that the response of the missile to a demand for rate of turn, $\dot{\psi}_m$, is uniform from 0 to B Hz ($< f_g$), the variance of the glint component of the rate of turn of the missile is given by

$$\sigma_{m^2} = (\lambda\,2\pi\sigma_g/R)^2\,(1/2f_g)\int_{-B}^{B} f^2\,df = (\lambda\,2\pi\sigma_g/R)^2\,(B^3/3f_g)\ (\text{rad s}^{-1})^2 \quad (6.9)$$

where λ is the navigation constant. As the missile closes on the target, the glint causes ever-increasing hunting until the missile is oscillating between the two extremes of maximum lateral acceleration. Postulate that this condition prevails when σ_m is equal to or greater than the maximum rate of turn of the missile $\dot{\psi}_m$ (max); from equation 6.9 this occurs when

$$R = (2\pi\sigma_g/\dot{\psi}_m(\text{max}))\,(B^3/3f_g)^{\frac{1}{2}} \quad (6.10)$$

Substituting the following typical values

$$\lambda = 5,\ \sigma_g = 4\text{ m},\ \dot{\psi}_m(\text{max}) = 0.15\text{ rad s}^{-1},\ B = 1\text{ Hz}, f_g = 2\text{ Hz}$$

gives

$$R = 340 \text{ m}$$

If the closing speed is 800 m s^{-1}, the time to impact, t, is 0.42 s. In this time, which is less than the missile response time, the missile will tend to turn at the maximum rate in one direction only. If the missile velocity is u_m, the lateral acceleration is $u_m \dot\psi_m(\text{max})$, so that the lateral displacement, d, is $u_m \dot\psi_m(\text{max})t^2/2$. Assuming that the missile was following a constant bearing terminal course, there would be no external demand and d would be wholly miss distance. Taking u_m as 600 m s^{-1} gives

$$d = 8 \text{ m due to glint}$$

which is a typical value. If the target is manoeuvring, so that there is a finite external demand, saturation occurs earlier and the miss distance due to glint is greater.

Multiple targets

Consider two identical targets, A and B, flying side-by-side at a velocity u_t as shown in figure 6.18. At first the missile sees only a composite target whose radar centre is at O and which is glinting strongly. As the missile closes on the targets it will eventually discriminate against one and turn to home on the other. In order to hit, it must discriminate soon enough to be able to execute the necessary turn. Discrimination is on the basis either of angle or of differential velocity. (D. K. Barton[8] discusses the problem of discrimination in general.) The seeker can discriminate in angle if the angle subtended by the two targets $2\delta\theta > \theta_3$.

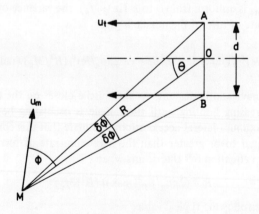

Figure 6.18   *Multiple targets*

Assuming $d \ll R$, $2\delta\theta = (d\cos\theta)/R$ so that discrimination takes place at a range $R = (d\cos\theta)\theta_3$. Obviously the most favourable conditions are when $\theta = 0$ or π. The guidance receiver can discriminate in velocity if the differential closing velocity $\delta u > \lambda B/2$, where λ is the wavelength and B is the width of the Doppler filter. The closing velocity of M and A is

$$u_m \cos(\phi - \delta\phi) + u_t \cos(\theta + \delta\phi)$$

and that of M and B is

$$u_m \cos(\phi + \delta\phi) + u_t \cos(\theta - \delta\phi)$$

The difference

$$\delta u = 2\,(u_m \sin\phi - u_t \sin\theta)\,\delta\phi; \delta\phi \to 0$$

Since

$$2\delta\phi = (d\cos\theta)/R, \delta u = \frac{d}{R}\left[u_m \sin\phi \cos\theta - \frac{u_t}{2}\right]\sin 2\theta \qquad (6.11)$$

the most favourable geometry is when $\phi = 0$; that is, when the missile is approaching directly. The maximum magnitude of δu for this case is $du_t/2R$, occurring at $\theta = \pi/4$, so that discrimination takes place at a range $du_t/B\lambda$. To give an idea of magnitude of the two means of discrimination, assume $d = 50$ m, $\theta_3 = 0.07$ rad, $u_t = 660$ m s^{-1}, $\lambda = 22$ mm, and $B = 500$ Hz. Angular discrimination is possible at a maximum range of $50/0.07 = 700$ m; velocity discrimination is possible at a maximum range of

$$\frac{50 \times 660}{22 \times 500 \times 10^{-3}} = 3\text{ km}$$

Although this indicates the superiority of velocity discrimination, the respective maximum values occur at different angles of approach and figure 6.19 shows this variation of discrimination range with angle of approach.

In practice, the missile will be flying a proportional navigation course so that it never flies directly towards the target. The graph for the velocity discrimination range in this example for a squint angle, ϕ, of 20° shows a peak value of 1 km at $\theta = 58°$, zero at 31°, but increasing sharply for smaller values of θ; this supports the contention that velocity discrimination is generally more useful than angle discrimination. If the missile uses a phase lock loop with B, say, 100 Hz, velocity discrimination will be possible at several kilometres.

Image effects[9]

These are most likely to be of importance when a low-flying target (figure 6.20) is being engaged. The seeker sees two targets: the real one, T, and its image, I, caused by specular reflection from the earth's surface at S. Serious difficulty can

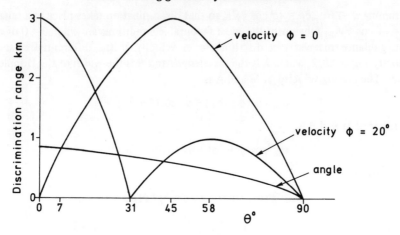

Figure 6.19   *Discrimination ranges*

arise if I approaches T in magnitude. This is most likely to occur over water, where the specular reflection coefficient tends to be high, approaching unity for a smooth surface at grazing incidence. T and I constitute a pair of targets, similar to those described under 'multiple targets', and the problems and techniques of discrimination are the same. There is one important difference; with multiple targets, it may not matter which one the missile engages eventually, but it is obviously vital that it engages the genuine target and not fly towards its image into the sea. Various techniques to distinguish the target from its image exist[10]

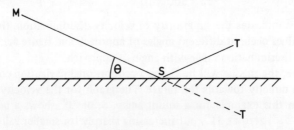

Figure 6.20   *Target image*

but they are no more than palliatives and some could not be fitted within the confines of a missile. The most valuable measure is to maintain I small in comparison with T; that is, to reduce the reflection coefficient. To this end, vertical polarization is desirable, and the wavelength should be as short as possible so that the sea surface appears rough; in addition the missile should, if possible, dive at the Brewster angle, which is about $7°$ at microwaves, in order to benefit from the reflection coefficient minimum.

6.12 Electronic counter-measures (ECM and ECCM)

There is a vast open literature on radar ECM and ECCM, though, naturally, applications to specific systems and quantitative assessments of effectiveness remain, for the most part classified. The ECM Handbook[11] is a useful compendium, Grant and Collins[12] give an up-to-date review and Johnson[13] gives a more specialized account of the pertinence to radar guided missiles. The purpose of this section is to describe the use of ECM against radar seekers for airborne targets, and the corresponding ECCM. ECM against surface target radar homers are described later, in chapter 8. The description does not venture into the more general topic of ECM against supporting radar and communication links since this is covered in more general texts.

Chaff[14]

This presents only a limited threat to air defence radar seekers since they tend to reject naturally the slow-moving chaff in favour of a fast-moving target. It would require a chaff echo of several thousand square metres to overcome the clutter rejection capabilities of a coherent seeker receiver sufficiently to obscure a small, fast, airborne target. Chaff can, however, be more effective if the target is crossing at the time of acquisition, since the target has no component of velocity towards the missile and the chaff Doppler frequency may fall within the search ambit of the Doppler frequency gate. Even under these circumstances it may still be possible to reject the chaff echo on the basis of range as the chaff separates from the target, by using pulse-Doppler; the seeker would discriminate between the target and the chaff, each in its own range interval, by the broader frequency spectrum of the wind-blown chaff.

Decoys

For the same reasons, slow-moving passive decoy reflectors are of limited value as an ECM, though a decoy with rotating vanes may be effective against a seeker with a conical scan or sequential lobing angle tracking system.

Confusion ECM, noise jamming

Once having detected an illuminating radar signal by means of an intercept receiver, the jammer radiates a noise-like signal over a fairly narrow band centred on the radar radio-frequency. Jamming ERP density (P_j) is expressed in watts per MHz. Jammers are classified as self-screening (SSJ), stand-off (SOJ) and ground-based (GBJ). SSJ are carried by attacking aircraft and constitute a small part of the payload. A SOJ aircraft is laden with ECM equipment and stands off out of range of the missile defences, generating sufficient power to provide a jamming umbrella for the waves of attacking aircraft. Unlike the SSJ, it is probably not within the main lobe of a missile directed against an attacking air-craft, hence it must generate sufficient power to be effective through a side-lobe of the seeker antenna. The GBJ is a still higher power radiator, situated well outside the immediate combat area; the same proviso concerning seeker antenna side-lobe applies. It is possible to make a numerical estimate of the effectiveness of noise jamming by the use of range equations, as the following examples show. The jamming power J, received by a seeker is given by the equation

$$J = \frac{P_J B F A_M}{4\pi R_{JM}^2 L_J} \tag{6.12}$$

where B is the seeker receiver bandwidth, F is a factor representing whether jamming is into the seeker antenna main beam ($F = 1$) or into a side-lobe, R_{JM} is range jammer–missile, and L_J represents miscellaneous losses. The target skin echo received by an active seeker is given by

$$S = \frac{P_I G_I \sigma A_M}{(4\pi)^2 L R_{TM}^4} \quad \text{(from chapter 3, equation 3.6)} \tag{6.13}$$

and by a semi-active seeker

$$S = \frac{P_I G_I \sigma A_M}{(4\pi)^2 L R_{IT} R_{TM}^2} \quad \text{(from chapter 3, equation 3.7)} \tag{6.14}$$

where suffix TM indicates target–missile, and suffix IT illuminator–target.

Example 1. SSJ, active seeker; $R_{JM} = R_{TM}$, $F = 1$. From equations 6.12 and 6.13

$$S{:}J = \frac{P_I G_I}{P_{JB}} \times \frac{\sigma}{4\pi R_{TM}^2} \times \frac{L_J}{L} \tag{6.15}$$

Since the jamming signal is noise-like, the target is unmasked when $S{:}J = S{:}N_{min}$. Substitute into equation 6.15 the numerical data of the example of section 3.2

$$P_I = 50 \text{ W CW}$$

$$G_I/4\pi = A_M/\lambda^2 = 0.025/(22 \times 10^{-3})^2$$

$$\sigma = 1 \text{ m}^2$$

$$L = 4 \text{ dB}$$

$$S{:}N_{min} = +6 \text{ dB}$$

$$B = 1 \text{ kHz}$$

and assume that the jammer data are $P_J = 1 \text{ W MHz}^{-1}$, $L_J = 3 \text{ dB}$; then

$$R_{TM} = 720 \text{ m}$$

This contrasts sharply with $R_{max} = 12$ km for detection of the target under quiet conditions.

Example 2. SSJ, semi-active seeker; $R_{JM} = R_{TM}$, $F = 1$

$$S{:}J = \frac{P_I\,G_I}{P_{JB}} \times \frac{\sigma}{4\pi R_{IT}^2} \times \frac{L_J}{L} \tag{6.16}$$

Substitute the illuminator data of the example of section 3.3 (P_I = 500 W CW, G_I = 47 dB) and other data as in example 1; then

$$R_{IT} = 20 \text{ km}$$

compared with an acquisition range of 63 km under quiet conditions for 'all-the-way' homing. The calculation for R_{IT} is independent of the range R_{TM}, hence a terminal semi-active homer similarly cannot acquire the target until it is within 20 km of the illuminator. This demonstrates that, under severe jamming conditions, terminal semi-active homing offers no range advantage over 'all-the-way' homing. It is fairly easy to deal with the SSJ by switching to a passive home-on-jam mode; switching is either automatic or on command from the launch point. In this context, TVM semi-active homing has the advantage that the signal actually received by the seeker is under continual monitoring at the illuminator. Should the target cease jamming during the missile flight, the seeker velocity gate should be able to re-acquire the target readily by searching for the echo about the last known Doppler frequency. A more difficult situation arises if two adjacent targets, close enough to lie in the seeker antenna main beam together, jam alternately; the missile must then rely on early angular discrimination and on sufficient agility to turn towards the remaining target (see section 6.10).

Example 3. SOJ, active seeker

$$S{:}J = \frac{P_I\,G_I}{FP_J\cdot B} \times \frac{\sigma}{4\pi} \times \frac{R_{JM}^2}{R_{TM}^2} \times \frac{L_T}{L} \tag{6.17}$$

Assume that P_J = 100 W MHz^{-1}, R_{JM} remains sensibly constant at 30 km, and jamming is into side-lobes of average level -20 dB with respect to the main lobe so that F = 0.01; other data is as in example 1. Then

$$R_{TM} = 4.6 \text{ km; the SOJ is less effective than the SSJ}$$

Example 4. GBJ, semi-active seeker

$$S{:}J = \frac{P_I\,G_I}{FP_J\,B} \times \frac{\sigma}{4\pi} \times \frac{R_{JM}^2}{R_{IT}^2\,R_{TM}^2} \times \frac{L_T}{L} \tag{6.18}$$

Assume that P_J = 1 kW MHz^{-1}, R_{JM} is sensibly constant at 80 km, F = 0.01, and other data is as in example 2. Then

$$R_{IT} \times R_{TM} = 566 \text{ km}^2$$

The significance is that an 'all-the-way' homer will be unable to acquire the target until R_{TM} = 22.5 km, while, if the target remains at a fairly constant range of 63 km from the illuminator, a terminal phase homer could not obtain lock until it was within 8 km of the target. In practice, the figures could be worse because of the shortening of the range R_{JM} during missile flight. The ECCM consists of rejecting the interfering signal because it is arriving off the antenna main lobe axis. In a monopulse receiver each angle channel contains not one, but two velocity gates; these are adjacent and of the same width. The residual narrow band target difference signal will lie in only one gate, but the broader band jamming signal will lie in both. The gates are narrow compared to the bandwidth of the jamming, hence there will be strong temporal correlation between the jamming signals in the two gates; subtracting the outputs of the gates cancels the jamming, leaving the small on-axis target difference signal. Note that, if the target illumination is separate from the target tracking transmission, the jammer may be deceived into jamming the wrong frequency. The principle in a conical scan receiver is similar except that there is just one pair of velocity gates, in the single-channel receiver. Strong conical scan modulation is present equally on the jamming in the two gates; after separate video demodulation the two gate outputs are subtracted, leaving only the residual conical scan signal due to target misalignment.

Deception ECM

Spin frequency jamming is an obvious way of deceiving a seeker equipped with conical scan or sequential lobing. A semi-active seeker is relatively immune since it does not betray the lobing rate by radiating; an active seeker would be very vulnerable and the employment of monopulse is mandatory. Velocity gate pull-off consists of the jammer retransmitting a narrow band signal derived from the illumination frequency. The jamming sweeps over at least the whole Doppler frequency coverage of the seeker, dumping the velocity gate at an extremity. ECCM are a rapid re-acquisition circuit, rejection of excessive velocity rate, coded illumination, and use of TVM guidance.

References

1. Long, J. J. and Ivanov, A., 'Radar Guidance of Missiles', Paper III-3 in Barton, D. K. (ed.), *Radars, Vol. 7, CW and Doppler Radar*, Artech House, 1978.
2. Ivanov, A., 'Semi-active radar guidance', *Microwave Journal*, September, 1983, p. 105ff.
3. Blanchard, A., *Phase-lock Loops*, Wiley, New York, 1977.
4. Acker, A. E., 'Eliminating transmitted clutter in Doppler radar systems', *Microwave Journal*, November 1975, p. 47ff.
5. Skolnik, M. I., *Introduction to Radar Systems*, 2nd edn, McGraw-Hill, 1980, Section 5.5.

6. O'Hara, F. J. and Moore, G. M., 'A high performance cw receiver using feed-through nulling', *Microwave Journal*, Vol. VI, No. 9, p. 63ff.

7. Glazier, E. V. D. and Lamont, H. R. L., *Transmission and Propagation, Services Textbook of Radio Vol. 5*, HMSO, 1958, Appendix 14.3.

8. Barton, D. K. (ed.), *Radars, Vol. 4, CW and Doppler Radar*, Artech House, 1975, Section VII. Multiple Target Estimators.

9. Evans, G. C., in Barton, D. K. (ed.), *Radars, Vol. 4, CW and Doppler Radar*, Artech House, 1975, Section I. Introduction, Paper 1.5.

10. Barton, D. K. (ed.), *Radars, Vol. 4, CW and Doppler Radar*, Artech House, 1975, Section VI. Complex angle techniques.

11. Boyd, A. J., *et al., Electronic Countermeasures*, Peninsular Publications, 1978.

12. Grant, P. M. and Collins, J. H., 'Introduction to Electronic Warfare', *IEE Proc.*, Vol. 129, Part F, No. 3, June 1982, pp. 113-30.

13. Johnson, S. L., 'Guided missile ECM/ECCM', *Microwave Journal*, September 1978, p. 22ff.

14. Butters, B. C. F., 'Chaff', *IEE Proc.*, Vol. 129, Part F, No. 3, June 1982, pp. 197-201.

7

Pulse-Doppler

7.1 Introduction

Pulse-Doppler[1] has tended recently to replace pure CW in homing against airborne targets, particularly in AAM and in terminally guided SAM. The reasons for this are: firstly, pulsing the transmitter permits time sharing between transmission and reception, so avoiding difficulties of duplexing and of rear reference spill-over; secondly, it provides an extra coordinate for target discrimination – range – which is particularly useful in a high density of targets; and thirdly, illumination in semi-active AAM is compatible with the primary role of the aircraft's fire direction radar, which is to locate and track targets.

7.2 Choice of PRF[2]

Pulse-Doppler radars are classified as low, medium and high PRF.

Low PRF

These are basically pulse radar, which have the additional capability of detecting moving targets by the presence of the Doppler shift on the carrier frequency. The low PRF, which is a few kHz at most, ensures that range measurement is unambiguous. Unfortunately, it gives rise to many velocity ambiguities, so ruling it out for homing against airborne targets.

High PRF

This is basically chopped CW. The PRF is sufficiently high to avoid velocity ambiguities, that is greater than $2 \times$ highest f_D, making it typically a value of several 100 kHz. The duty cycle is high, up to 50 per cent. Range gating is hardly worth while, and the main purpose is to eliminate the difficulties of duplexing in active homing and the effects of rear reference spillover in semi-active homing.

Medium PRF

As the name implies, this is a compromise between low and high PRF to take advantage of the useful features of each. The main advantage is that it permits both range and velocity gating, providing the degree of discrimination needed in a high density of targets. The disadvantage is that the PRF, which typically has a value of a few 10s of kHz, gives rise to both range and velocity ambiguities. These ambiguities are resolved by the use of multiple PRFs, though at the cost of a longer processing time.

7.3 High PRF seekers

The guidance receiver is essentially CW, with Doppler frequency acquisition and tracking. The only addition is a blanking pulse to cut off the receiver for the duration of the transmitted pulse or of the reception of the rear reference pulse. In active homers the transmitter modulator supplies the blanking pulse; in semi-active homers it comes from the rear reference receiver. Figure 7.1 shows the arrangement for a semi-active homer: blanking is usually in the RF feeders but it can be applied also to the IF amplifier in order to cut out receiver noise during the blanking period. Unfortunately, blanking causes a loss of signal strength by the eclipsing of the received pulse by the blanking pulse; the higher the duty cycle the greater the loss, as figure 7.2 shows. Instantaneous loss depends upon the overlap of the train of pulses; this overlap varies as the range changes, causing a cyclic loss of signal strength. The cyclic rate, which is a few Hz, is equal to the Doppler shift of the PRF. The proof is as follows: one eclipsing cycle occurs during a change in elapsed time of T_r, equivalent to a change in range of $cT_r/2$. If the target velocity is u, this change occurs in a time of $cT_r/2u$, equivalent to

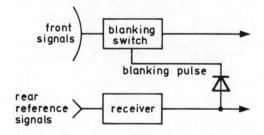

Figure 7.1   *Blanking in semi-active pulse-Doppler receiver*

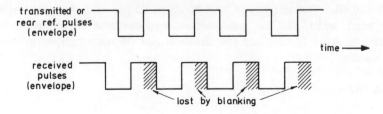

Figure 7.2   *Eclipsing – 50 per cent duty cycle*

an eclipsing rate of $2uf_r/c$, which is the Doppler shift on the PRF. The important factor is the loss of average signal: noise $(S:N)$ over one eclipsing cycle
(eclipsing loss). Formulae for eclipsing loss are derived below and table 7.1 sets
out specimen figures.

Table 7.1. Eclipsing loss (averaged over an eclipsing period)
(The table is of average signal: noise $(S:N)$ with respect to $S:N$ with no blanking
or range gating – dB.)

Duty cycle	Signal only blanked	Signal and noise blanked	Signal and noise blanked and range-gated
50 per cent (1:1)	−4.8	−1.8	−1.8
25 per cent (1:3)	−1.8	−0.5	+4.3
1:n (n large as in ordinary pulse radar)	0	0	$+10 \log_{10} n$

Received signal only blanked

The blanking is applied in the RF feeder from the antenna, cutting off the signal
but not affecting the receiver noise. The timing, t (figure 7.3), of the blanking
pulse relative to the signal pulse depends upon the range, and as the range changes

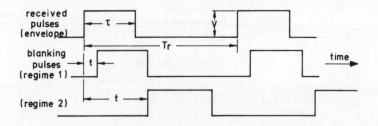

Figure 7.3   *Eclipsing regimes*

t covers all values from 0 to T_r in an eclipsing cycle. During at least part of the eclipsing cycle, some of the received signal pulse is lost by overlap of the blanking and signal pulses (eclipsing), as in regime 1. During regime 1, which occurs for values of t between $-\tau$ and τ, the unblanked portion of the signal pulse is t. During regime 2, which occurs for values of t between $(T_r - \tau)$ and τ, the unblanked portion is constant, of duration τ. The received pulses are integrated coherently so that the mean value of signal energy over an eclipsing cycle given by

$$\frac{V^2}{T_r}\left[\int_{-\tau}^{\tau} t^2\,dt + \tau^2 \int_{\tau}^{T_r-\tau} dt\right]$$

$$= \frac{V^2\tau^2}{T_r}\left[T_r - \frac{4\tau}{3}\right] \tag{7.1}$$

Without blanking it would be $V^2\tau^2$; hence the eclipsing loss, which is the ratio of the two expressions, is

$$\left[1 - \frac{4\pi}{3T_r}\right] \tag{7.2}$$

For $\tau = T_r/2$ (50 per cent duty cycle) the ratio is one-third — a loss of 4.8 dB. For 25 per cent duty cycle it is two-thirds, a loss of 1.8 dB; beyond this the loss becomes negligible as the duty ratio falls.

Signal and noise blanked

Let the mean noise power be N. If the noise is blanked its mean value is reduced by $N(1 - \tau/T_r)$, hence the net $S{:}N$ due to eclipsing is

$$(1 - 4\tau/3T_r)/(1 - \tau/T_r) \tag{7.3}$$

compared to an unblanked signal. At $\tau = T_r/2$ the loss is now only 1.8 dB. If the signal is also range-gated, the average noise power is reduced to $N\tau/T_r$, giving the net $S{:}N$ as

$$(1 - 4\tau/3T_r)\,(T_r/\tau) \tag{7.4}$$

compared to a signal which is neither blanked nor range gated.

At $\tau = T_r/2$ the loss remains at 1.8 dB, but at $\tau = T_r/4$ the factor is 8/3, or a gain of 4.3 dB. These figures confirm that range gating is hardly worth while at high duty cycles, but the lower the duty cycle the more worth while it becomes. In practice, it would be applied if the duty cycle were about 0.1 or less.

Provision of the reference signal

As this must be available at all times, it must be CW. In an active seeker CW, reference is available readily from the transmitter master oscillator. In a semi-

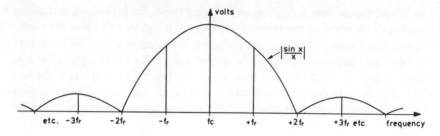

Figure 7.4   *Frequency spectrum of train of rectangular pulses – 50 per cent duty cycle*

active seeker, the rear reference is pulsed and the main line of the spectrum (figure 7.4) must be selected by means of a narrow bandpass filter of width no more than $2f_r$.

7.4 Medium PRF seekers

These provide resolution in both range and velocity by means of tracking gates working in tandem. Figure 7.5 shows a possible arrangement. Typical resolutions are a few hundred metres in range and a few hundred hertz in Doppler frequency.

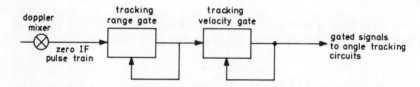

Figure 7.5   *Tracking in both range and velocity (explicit Doppler receiver)*

It is possible to employ conventional analogue methods to resolve targets but the advent of fast microcomputers makes it more convenient to use digital methods, which are both more flexible and more compact in apparatus.

Acquisition

Since these are essentially Doppler systems, it is necessary to acquire the target after launch for the reasons explained already in chapter 6. If it is an 'all-the-way' homer, the estimated target range and velocity are stored in the guidance receiver before launch; if it is a terminal homer, the parent guidance radar transmits the target range and velocity to the missile in flight. In either case the guidance receiver has to search for the target in both range and velocity. Because of the consequent large number of resolution cells, it is desirable to use searching in parallel by means of a bank of range gates, each with its own bank of velocity filters. For example, a range search over 1.5 km with five range gates each of 300 m, and a Doppler frequency search over 5 kHz with ten filters each of a bandwidth of 500 Hz, covers 50 resolution cells.

An analogue system would use LC bandpass filters; these are fairly bulky and it is not easy to change characteristics such as bandwidth. Digital filtering offers a more compact and more flexible means of signal processing. Because of their novelty, the digital methods are described briefly later in this section.

Resolution of ambiguity

As explained in section 7.2, medium PRF pulse-Doppler radar suffers from both range and velocity ambiguities and the acquisition and tracking systems must be capable of resolving them. The established method is by the use of multiple PRF[3], as follows. If there are Doppler frequency ambiguities, the Doppler frequency, f_D, is given by

$$f_D = f_{Do} + nf_r \qquad (7.5)$$

where f_{Do} is the observed Doppler frequency $(< f_r)$, f_r is the PRF, and n is an integer. Since n is unknown, the estimate of f_D is ambiguous. Similarly, if there are elapsed time (range) ambiguities, the elapsed time, T_e, is

$$T_e = T_{eo} + k/f_r \qquad (7.6)$$

where T_{eo} is the observed elapsed time $(< 1/f_r)$, and k is another integer. If two PRFs, f_{r1} and f_{r2} $(f_{r1} > f_{r2})$, are used there is no ambiguous f_D common to both below a frequency $f_{r1} \times f_{r2}/(f_{r1} - f_{r2})$. By comparing the two values of f_{Do}, the ambiguity in Doppler frequency can be resolved for all frequencies below this value. Similarly, elapsed time ambiguities can be resolved for an elapsed time of less than $1/(f_{r1} - f_{r2})$. Say that the maximum Doppler frequency for the missile is f_{Dmax} and the maximum elapsed time is T_{emax}; then

$$\frac{(f_{r1} \times f_{r2})}{f_{r1} - f_{r2}} \geqslant f_{Dmax} \qquad (7.7)$$

$$\frac{1}{f_{r1} - f_{r2}} \geqslant T_{emax} \qquad (7.8)$$

The easiest way of dealing with this pair of inequalities is to take one of them as an equality and substitute in the other one. For example, taking 7.8 as an equality, and substituting for f_{r1} in 7.7 yields a quadratic inequality for f_{r2} whose valid solution is

$$f_{r2} \geqslant \frac{(4T_{emax} \times f_{Dmax} + 1)^{\frac{1}{2}} - 1}{2T_{emax}} \qquad (7.9)$$

which gives the corresponding inequality

$$f_{r1} \geqslant \frac{(4T_{emax} \times f_{Dmax} + 1)^{\frac{1}{2}} + 1}{2T_{emax}} \qquad (7.10)$$

Taking $T_{emax} = 80 \ \mu s \ (\equiv 12 \ km$ of range) and $f_{Dmax} = 120 \ kHz$

$$f_{r2} \geqslant \frac{(39.4)^{\frac{1}{2}} - 1}{0.16}; \ f_{r1} \geqslant \frac{(39.4)^{\frac{1}{2}} + 1}{0.16}$$

For practical reasons it is desirable that the two PRFs should have a simple integral relationship, hence the number under the square root sign should be a perfect square, greater than the existing number. In this example the next highest perfect square is 49, which yields

$$f_{r2} = 37.5 \ kHz; \ f_{r1} = 50 \ kHz$$

which are in a ratio of 3.4.

These correspond to $f_{Dmax} = 150 \ kHz$, $T_{emax} = 80 \ \mu s$, thus satisfying the requirements. Equally 7.7 could have been taken as the equality, yielding a different pair of PRFs, which would still have satisfied the requirements. With a pair of PRFs, the radar obtains an unambiguous value of f_D by applying equation 7.5 to each observed value separately, looking for a pair of values of the integer n which gives a common value of f_D. For values of $f_d < f_{Dmax}$ the only valid pairs of n are those which are equal or which differ by unity. Say $f_{Do1} = 30 \ kHz$ and $f_{Do2} = 5 \ kHz$: with the above pair of PRFs the lowest common value of f_D is 80 kHz, corresponding to $n = 1$ and 2 respectively. If two or more targets are present there will be a value of f_{Do} for each target and for each PRF, and the radar will then have to examine every possible pairing for a valid common f_D. Say that $f_{Do1} = 10 \ kHz$ and $f_{Do2} = 35 \ kHz$ are present also, then the following pairings are possible

f_{Do1}	f_{Do2}	n_1	n_2	f_D
30	5	1	2	80
30	35	no integral values		–
10	5	no integral values		–
10	25	2	2	110

thus resolving the ambiguities. Note that the number of possible pairings goes up as the square of the number of targets, which can include clutter under unfavourable circumstances, so that sorting the correct pairs can be a lengthy process. With only two PRFs, there are regions of low sensitivity at Doppler frequencies which are a multiple of either PRF and at elapsed times which are a multiple of either recurrence period. Use of more PRFs leads to a more uniform response, at the price of a computation time which is lengthened by the factor $^xC_2$ (combination of x-things taken 2 at a time), where x is the number of PRFs.

7.5 Radar digital signal processing

The large amount of data to be processed in a pulse-Doppler radar suggests the desirability of using digital methods. Digital signal processing is described briefly by Skolnik[4], and in more detail by Sheats[5] and in Oppenheim[6]. Its advantages compared with analogue signal processing are that it is performed by a number of small and cheap digital logic components, that the stability is good, and that the processing is flexible. The disadvantage is that it is not yet fast enough to deal with all radar signal processing requirements; however, the speed is now sufficient for most purposes, including those connected with radar homing missiles, and furthermore the state of the art advances continuously, often quite rapidly.

The following is an account of the digital treatment of pulse-Doppler signals. The output of the Doppler mixer is a train of zero-IF or 'bipolar video' pulses, as shown in figure 7.6. They are first of all gated by a bank of fixed range (or time) gates. The signals in each gate are then digitized for analysis in Doppler frequency.

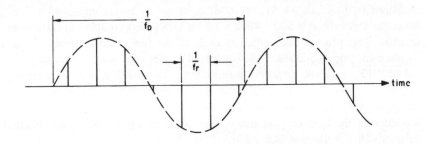

Figure 7.6   *Train of bipolar video pulses*

The digital sampling rate must be at least once per pulse duration and the number of quantization levels must be sufficient to ensure that quantization noise does not limit the inherent clutter rejection capability. The inherent clutter rejection capability of pulse-Doppler radar is discussed by Skolnik[1] and can be expected to be about 40–50 dB. Accordingly, a quantization noise level of −60 dB would be acceptable, indicating the need for about 10 levels[7]. The frequency analysis of the train of digital pulses is carried out by taking the Discrete Fourier Transform (DFT), whose operation is as follows. Let $f(t)$ be a continuous periodic function of time, of period T. Let it be sampled at a rate of

N times in the interval T; the sampled signal can be represented as $f(r)$, where $r = tN/T$ and is an integer lying between 0 and $N - 1$. The Fourier analysis of the sampled waveform is

$$F_k = \frac{2}{N} \sum_{r=0}^{N-1} f(r)\exp(-j2\pi kr/N) \tag{7.11}$$

where k is an integer lying between 0 and $N - 1$.

To evaluate each term, F_k requires N complex multiplications, hence evaluation of the complete spectrum requires N^2 multiplications. However, many of the calculations are repeated because of the cyclic nature of the exponential terms. By eliminating the repetitions by an algorithm known as the Fast Fourier Transform (FFT), it is possible to reduce the total number of complex multiplications to $N \times \log_2 (N)$, thus reducing considerably the computation time. The FFT is described in several works[5,6,8] but in view of its central importance a brief account is given in appendix C. As an example of digital signal processing using the FFT, consider again the initial acquisition of a target by the pulse-Doppler homer mentioned in section 7.4. After range-gating, the waveform to be sampled is the Doppler frequency. Say that the highest Doppler frequency is 120 kHz, then the sampling rate must therefore be 240 kHz. The observation time must be long enough to provide a Doppler frequency resolution of 500 Hz – that is, 2 ms – hence the number of samples per observation period is 480. To implement the FFT it is desirable that this number should be an integral power of 2, therefore choose $N = 512$, that is 2^9. The number of complex multiplications is therefore $9 \times 512 = 4608$; these are to be accomplished in 2 ms, hence the throughput is $500 \times 4608 = 2.3$ MCOPS (mega complex operations per second). This process takes place in each of the five range intervals, giving a requirement for a 5-channel, 512-point FFT processor with a throughput of 2.3 MCOPS. A review article[9] confirms that this is within current capability. This FFT processor could handle a CW signal equally well; all that is necessary is to bypass the range gates. Hence a missile equipped with this pulse-Doppler processor in the homing guidance system could be used with either coherent pulse or with CW illumination.

References

1. Skolnik, M. I., *Introduction to Radar Systems*, 2nd edn, McGraw-Hill, 1980, Chapter 4.
2. Aronoff, E. and Greenblatt, N. M., 'Medium prf design and performance', Paper IV-7 in Barton, D. K. (ed.), *Radars, Vol. 7, CW and Doppler Radar*, Artech House, 1978.
3. Skilling, W. A. and Mooney, D. H., 'Multiple high prf ranging', Paper IV-2 in Barton, D. K. (ed.), *Radars, Vol. 7, CW and Doppler Radar*, Artech House, 1978.

4. Skolnik, M. I., *Introduction to Radar Systems*, 2nd edn, McGraw-Hill, 1980, Section 4.5.
5. Sheats, L., 'Fast Fourier Transform' (chapter 9), in Brookner, E. (ed.), *Radar Technology*, Artech House, 1977.
6. Oppenheim, A. V. (ed.), *Application of Digital Signal Processing*, Prentice-Hall, 1978, Chapter 5.
7. Skolnik, M. I., *Introduction to Radar Systems*, 2nd edn, McGraw-Hill, 1980, Eqn. 4.12.
8. Brigham, E. O., *The Fast Fourier Transform*, Prentice-Hall, 1974.
9. Brookner, E., 'Developments in digital radar processing' and 'A guide to radar systems and processors', in Oppenheim, B. V., (ed.), *Trends and Perspectives in Signal Processing*, Vol. 2, No. 1, January 1982, pp. 7.2ff; revised and condensed as 'Trends in radar signal processing', *Microwave Journal*, October 1982, p. 20ff.

8

Radar Homing on Surface Targets

8.1 Introduction

Radar homing on surface targets is possible only if they can be seen against the background of surface clutter. Until recently this has been a practical proposition only against ships; a ship presents a large target and sea clutter is generally less than land clutter; it is also less well correlated. However, by the use of mm waves, seekers are now available with sufficiently good angular resolution to pick out a large battlefield target, such as an armoured fighting vehicle (AFV) against the background of clutter.

Types of transmission

Surface targets move too slowly for reliable discrimination against clutter by the use of the Doppler effect; indeed, at sea this difficulty is compounded by the wide Doppler frequency spectrum of sea clutter arising from movement of the waves. CW-Doppler offers no advantage and better clutter rejection is obtained by pulse transmission, relying for discrimination on a small resolution cell formed of narrow beamwidths and a short pulse. Pulse-Doppler is, however, useful for improving angular resolution by the technique of 'synthetic aperture' or Doppler beam sharpening, as used in airborne surveillance by sideways-looking radar.

8.2 Sea targets

Most modern anti-ship missiles are sea skimmers; that is they approach the target at a constant height of a few metres above the sea. This makes the missile a difficult target to detect. The missile is usually launched from a platform below the horizon from the target; it then follows a mid-course trajectory under, say, inertial guidance until it comes within terminal homing radar range of the target. During the mid-course phase the missile will not necessarily have been at sea level; it will probably descend in steps to the final height during the homing radar acquisition phase and the early part of the tracking phase. Once the guidance radar has acquired the target, it tracks it in azimuth and range, generating the steering commands for the missile through the control system. Height above the sea is controlled separately by radar, or possibly laser, altimeter. Because of the remoteness of the launching platform the homing is usually fully active, though semi-active is possible if the launching platform is airborne at sufficient height to keep the target in view. The final height depends upon the sea state but it is as low as possible, just clearing the tops of the waves. The relationship between wave height and sea state appears in a useful table reproduced in ref. 1; for example, the wave height (crest to trough) at sea state 4 is 4 ft. A missile would have to fly at a height of at least three times this above the mean level — that is, at about 4 m to be reasonably certain of clearing the waves.

More conventional trajectories

Earlier missiles, such as Styx, attack along a more conventional diving trajectory; this requires angle tracking in both azimuth and elevation. A missile on such a trajectory is easier to detect; furthermore it may have to face more severe sea clutter.

Pulse duration

This must be as short as possible to reduce the area of clutter in the resolution cell. However, it must be long enough to straddle the target, otherwise signal strength is lost as a result of reduction of the area of the target illuminated. For example, say that the maximum beam of the vessel is 24 m. To be sure of enclosing the whole of this in the length of the resolution cell, the pulse length must be not less than $0.16\,\mu s$.

Factors affecting acquisition of targets

The range at which a target can be acquired depends upon the following external factors: the size of the target and its aspects, whether or not it is partially below the horizon, multipath effects and sea clutter. Since sea skimmers and more steeply diving missiles face different problems, it is better to consider the two separately. The sea skimmer sees the target at practically normal incidence. It is not easy to quantify the strength of a ship echo but Skolnik[2] gives an approximate formula for the median value of the echoing area of a fairly large steel-hulled ship, averaged over all quarters. It is $52\, f^{\frac{1}{2}}\, D^{\frac{3}{2}}\, m^2$, where f = frequency in MHz, and D is the fully laden displacement in kilotons. For example, a ship of 4000 tons will have a median echoing area of 4000 m^2 at 10 GHz. There is, however, a large variation of echoing area with horizontal aspect; variations of 20 dB are typical[3]. Another important consideration is the radar horizon; for low heights the standard distance is $4130\, h^{\frac{1}{2}}$ metres, where h is the height of the antenna above the sea; for example, for h = 4 m the horizon distance is about 8 km. Nevertheless, the seeker can see beyond the horizon because of the height of the ship; if the height of the hull were 9 m, the seeker could just see the top of the hull at a range of $4130(4^{\frac{1}{2}} + 9^{\frac{1}{2}})$, or about 20 km. However, the echo would be very small and a more reasonable estimate of visible range is that at which the hull echo is half its fully visible value. The echo from a large flat plate is proportional to the square of the area; assuming that the hull behaves as such a plate, the echo is proportional to the square of the height of the hull above the horizon. In this example the echo would be half its fully visible value when $9/2^{\frac{1}{2}}$ = 6.4 m of the hull height were exposed; hence 2.6 m are below the horizon, giving an additional horizon distance of $4130 \times 2.6^{\frac{1}{2}}$ = 6.7 km. Hence the visible range would be about 15 km.

Another possible difficulty arises from multipath effects caused by reflections from the surface of the sea. Because of the small grazing angle, reflections are likely to be strong, even at quite high sea states. According to plane earth reflection theory, the voltage reflection factor, one-way, is $2\sin(2\pi h_1 h_2/\lambda R)$ for complete reflection[4], where h_1 and h_2 are the radar and target heights respectively and R is the range. The target echo falls to its free-space value where the argument of the sine = $\pi/6$ — that is, at $R = 12 h_1 h_2/\lambda$; beyond this the target signal strength decays rapidly, according to R^{-8}. It is rather difficult to specify h_2 for a target on the surface of the sea but treating the target, as before, as a flat plate equal in area to the projected area of the hull, h_2 = height of the hull. Taking h_1 = 4 m, h_2 = 9 m, and λ = 32 mm gives the limiting range as 13.5 km. The combined effect of the horizon and of reflection would be to give a maximum visible range of somewhat less than 13.5 km in this case. The third effect is clutter; because of the small grazing angle, clutter is unlikely to pose difficulties except at sea states so high that the missile could not fly in any case. To show this, consider an active seeker, antenna horizontal beamwidth (θ_3) 80 mrad, range 12 km, and range resolution (ΔR) 30 m. Figure 8.1 depicts the resolution cell on the surface of the sea, together with the target T. The area of the cell is 29 000 m^2. At sea state 4 and frequency 9.4 GHz, the clutter echoing area per m^2 of sea surface, σ_0, can be taken as -48 dB for both horizontal and vertical polarization[5], giving a total clutter echoing area of 0.46 m^2, which is likely to be negligible in comparison with the echoing area of a ship.

A diving missile faces a different situation. The echoing area of the target obviously depends upon the vertical aspect, but is otherwise difficult to specify.

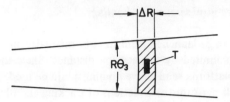

Figure 8.1   *Surface resolution cell*

An estimate is that it is at least 10 dB less than that seen by a sea skimmer approaching from the same quarter. On the other hand, the horizon is unlikely to set an operational limit and multipath effects are less important because of the small reflection coefficient at a large grazing angle. By the same token, clutter may be more important because of the greater back scatter. Referring to figure 8.2, area of the resolution cell is $R\theta_3 \Delta R \sec \phi$. Using the same numerical data and taking $\phi = 45°$ yields an area of $41\,000$ m^2; the value of σ_0 is now much larger, being -24 dB for vertical and -33 dB for horizontal polarization[5]. The corresponding clutter echoing areas are 163 m^2 and 21 m^2, indicating possible difficulty in detecting small ships, or even in detecting quite large ships at unfavourable aspect angles. This example also brings out the desirability of using horizontal polarization.

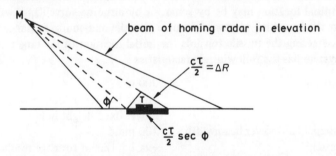

Figure 8.2   *Resolution cell – non-zero grazing angle*

Choice of frequency

The frequency band 9–16 GHz is the one used most commonly for radar homers on to sea targets. It has the advantage of fairly good all-weather performance, and the wavelength is not too long for adequate angular resolution. A higher frequency is attractive in giving a correspondingly finer antenna beam but sea clutter levels are higher and this may offset the benefit of the finer resolution cell. In addition, rain and spray attenuation may become unacceptable. The matter is still open and a designer must judge his case on its merits.

Acquisition – 'all-the-way' semi-active homing

The missile is likely to be launched from an airborne platform in order to give an adequate range, not limited by the horizon distance. Since the target is visible from the airborne platform, semi-active homing could be used. To economize on space, one radar will perform the functions of tracking and illuminating. Having acquired the target, the radar passes target coordinates, azimuth, and range to the missile on the launcher. The seeker acquires the target and when the *S:N* is strong enough the missile is launched. There is not the same difficulty in holding lock through boost as in a Doppler system; the range rate is quite slow so that the range tracking servo system has no difficulty in following. The only proviso is that signal strength should be sufficient to override the noise induced in the receiver by the vibrations of the boost phase. This vibration noise level can be determined only by experiment, but a typical value is + 10 dB with respect to the normal receiver noise.

Acquisition, terminal phase homing

A missile with terminal phase homing is launched either from over the horizon, or at least from well beyond the range of the target's own weapons. In either case it will be necessary for the terminal homing guidance system to search for the target at the commencement of the terminal guidance phase. Take as an example a sea-skimming missile launched from a surface vessel from over the horizon. Initial location may be by sonar, or by airborne surveillance with data relayed to the launching platform. Assume that the mid-course guidance system is inertial, steering the missile towards the initial location of the target. Assume that the system has the following characteristics

Missile velocity	300 m s^{-1}
Seeker	Pulsed radar, tracking in azimuth and range, fixed height hold
Seeker antenna half-power beamwidth	100 mrad
Pulse duration	$\frac{1}{3}\mu$s, equivalent to range resolution of 50 m
Terminal guidance phase	10 km
Probability of acquisition demanded	95 per cent
Errors of initial location and mid-course guidance, combined	200 m (1σ), both in longitudinal and transverse directions

The target is capable of a speed of 10 m s^{-1} (2 knots)

Assume that the missile is launched at a range of 40 km. The mid-course guidance phase is 30 km and lasts 100 s. In this time the target could have travelled up to 1 km in any direction. For 95 per cent probability of acquisition (PA) overall, PA in each coordinate must be 97.5 per cent so that the search must extend over ± 2.2σ (from tables of the normal distribution) – that is, over ± 440 m. To this must be added 1 km for the possible movement of the target. Hence the search in each coordinate must cover ± 1440 m. In azimuth this subtends ± 0.144 rad at 10 km, hence the total azimuth search should cover 3 beamwidths. In range it covers $2 \times 1440/50 = 58$ range intervals in all. Hence the total search covers $3 \times 58 = 174$ resolution cells. If it takes 20 ms to examine a resolution cell, the maximum search time is 3.5 s. Note that in this time the range to the target would have shortened to 9 km, increasing the subtended angle of the azimuth search to ± 0.16 rad; it might be desirable, therefore, to use a slightly wider search of 3.3 beamwidths overall.

8.3 Land targets[6,7]

These comprise tanks and other large vehicles, artillery etc. There is a requirement for a weapon, such as the US 'Wasp' and 'Assault Breaker', which can attack individual targets of formations stationed well to the rear and out of range of front line weapons[8]. Attack from the air by a missile which homes on to the target is a possible way of meeting this requirement. A radar solution would use an active pulse seeker which would lock on to the target after release and guide the missile towards impact. The missile could be delivered from an aircraft or it could be launched from the ground at some distance from the target as an inertially guided or free-flight rocket, or as an artillery shell; on reaching the vicinity of the target, the missile would enter the terminal homing phase.

The objection to radar homing on land targets has always been the difficulty of distinguishing the target against the background of land clutter. If the target is moving, it may be possible to distinguish it against a background of clutter by observing its Doppler frequency relative to that of the clutter (the 'clutter reference' or 'externally coherent' Doppler technique). However the target may well be stationary, or moving so slowly as to be indistinguishable from wind-blown clutter, hence the only sure way of seeing a target is to use a radar of sufficiently fine resolution. Resolution in range is not a serious problem, as pulse duration comparable with the target's dimensions can be generated. Resolution in angle is more difficult, particularly since the projectiles concerned tend to be small (6 inches or less in diameter). To obtain adequate resolution with such a small diameter it is necessary to use much shorter wavelengths than is customary in radar, in the millimetre wave bands. It is fortunate that rapid developments in mm wave technology in recent years have made small seekers a feasible proposition[9].

The frequency bands of interest are 35 GHz (8 mm), 94 GHz (3.2 mm), 140 GHz (2.1 mm), and possibly 220 GHz (1.4 mm), all of which lie in windows in the atmospheric attenuation characteristic. The attenuation is still relatively high, increasing with frequency, but it is acceptable at the short ranges envisaged. As an example of the resolution and of the signal:clutter power levels achievable, consider a fully active pulse homer at 94 GHz with an antenna diameter of 120 mm. Using an approximate formula, $\theta_3 = 1.5\ \lambda/d$ gives $\theta_3 = 0.04$ rad. The area of resolution cell on the ground is given by $(c\ \tau/2)\sec\phi \times R \times \theta_3$ as in figure 8.2. Assuming a pulse packet of length 10 m, $\phi = 30°$ and $R = 1$ km gives an area of 225 m^2. Taking $\sigma_0 = -20$ dB as a typical value for agricultural land at this grazing angle[10] gives a clutter echo of 2.25 m^2. The echoing area of a tank is likely to lie between 10 and 100 m^2, hence it should be visible at this range. At the longer wavelength of 8 mm the corresponding clutter echoing area would be 6 m^2, which could obscure the target. At 2 km range the resolution cell would be four times larger in area, hence the clutter power would be four times greater; this might create difficulty in detecting a target.

Precision guidance by synthetic aperture radar[11]

The calculations given above show that even mm wavelengths cannot always provide a narrow enough antenna beam for adequate resolution of land targets. It is possible to obtain much finer angular resolution than a beamwidth by the form of Doppler signal processing known as 'synthetic aperture'. This is well established already as a technique for high-resolution terrain mapping, using sideways-looking radar with a fan beam. What is required here, though, is a spotlight which can pick out individual targets within an already fairly narrow pencil beam. Figure 8.3 shows an active radar homing missile M with the radar beam pointing towards the ground, creating a fairly large 'footprint'. Longitudinal

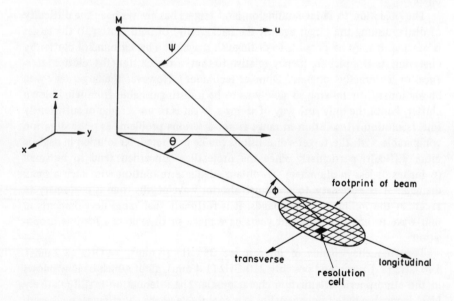

Figure 8.3   *Spotlight mapping by Doppler beam sharpening*

or range resolution within this footprint is given, as shown in figure 8.2, by $(c\tau/2)\sec\phi$. Assume that M is flying horizontally at a constant velocity u and that the beam is squinting at an angle ψ. Since the Doppler frequency of a stationary object within the beam depends upon ψ, it is possible to obtain transverse resolution by resolving the Doppler frequencies within the echo of the footprint. At the centre of the beam the Doppler frequency is

$$f_D = (2u/\lambda)\cos\psi$$

From the geometry

$$f_D = (2u/\lambda)\cos\theta \times \cos\phi$$

In the transverse direction ϕ is constant, hence f_D depends only upon θ. Therefore the incremental change in f_D is given by

$$\delta f_D = -(2u/\lambda)\cos\phi \times \sin\theta \times \delta\theta$$

δf_D is resolved by a bank of narrow bandpass filters spanning the total variation in f_D across the footprint; the resolution is the bandwidth B of each filter. Hence the corresponding resolution in θ is

$$\delta\theta = (\lambda/2u)\sec\phi \times \mathrm{cosec}\,\theta \times B$$

The transverse resolution at range R is $R\cos\phi\delta\theta$ which becomes

$$(\lambda RB/2u)\mathrm{cosec}\,\theta$$

Stipulating that the transverse and longitudinal resolutions shall be equal at the acquisition range R gives an expression for B

$$B = (2u/\lambda)(c\tau/2)(\sin\theta \times \sec\phi)/R \qquad (8.1)$$

The optimum value for θ is $90°$, as in sideways-looking radar, but in the present instance the beam must look forward, giving the missile time to turn and descend upon a target. As an example, choose the following quantities additional to those used in the first example of this section: $u = 300$ m s^{-1} and $\theta = 60°$. This gives

$$B = \frac{2 \times 300}{3.2 \times 10^{-3}} \times \frac{10}{2000} \times \sin 60° \times \sec 30°$$

$B = 700$ Hz, which is well within current capabilities

There remains the possible difficulty of range and Doppler frequency ambiguities. At 94 GHz, f_D at the centre of the beam is 81 kHz at these values. Since at 2 km range the maximum PRF for no ambiguity is 75 kHz, this would lead to ambiguities in the Doppler frequency. This difficulty disappears at the lower frequency of 35 GHz, although the process entails reducing B in proportion, to 260 Hz, a value which is still practicable.

Imaging

The figures quoted for resolution in the example above are by no means at the limit of current technology. Finer resolution is possible both in range and in angle (Doppler frequency), so that a large target could fill several resolution cells. This creates a radar image of the target which may be clear enough to provide a measure of identification. However, such fine resolution calls for exceptionally good frequency stability and spectral purity, and for exceptionally good control of missile speed and course.

Terminally guided sub-munitions (TGSM)[6,7,12]

Instead of launching a missile at a single target, it is possible to engage several targets of a group by means of a large missile conveying several small missiles, or sub-munitions, to the vicinity of the target area. Each sub-munition carries an active pulse seeker which acquires and homes on to one of the targets. These sub-munitions are necessarily small, typically 100 mm diameter and 600 mm long, total weight 7 kg, so that only a low-power transmitter is possible, restricting the homing range to a few hundred metres. The frequency is likely to be high, 94 GHz or above, in order to provide adequate angular resolution. The weight and space penalties attaching to a moving antenna can be avoided by using a fixed antenna, facing forwards, and spinning the sub-munition; the resulting conical scan signals provide the necessary pointing information.

Millimetre wave technology[6,8]

The antenna is likely to be a conventional paraboloidal reflector (figure 8.4a), though flat plate antennas (figure 8.4b) are under development[13]. Longer-range missiles operating at the lower frequency of 35 GHz may use conventional valve sources, such as the magnetron or some variant of the klystron. The higher-frequency, small, short-range systems will almost certainly use solid state sources, such as the IMPATT diode, which are now capable of delivering sufficient power for a range of 1-2 km. There has also been much improvement in the noise performance of mm wave receivers and, although no low-noise pre-amplifiers are in existence, noise figures are now as good as those of low-frequency receivers without pre-amplifier.

8.4 Radiometer or passive homing[14,15,16,17]

All the systems described above have been active, relying upon an illuminator, thus laying themselves open to active counter-measures. At the short mm wavelengths proposed already for homing on to surface targets, the natural thermal radiation from targets at terrestrial temperatures (around 290K) is sufficient to

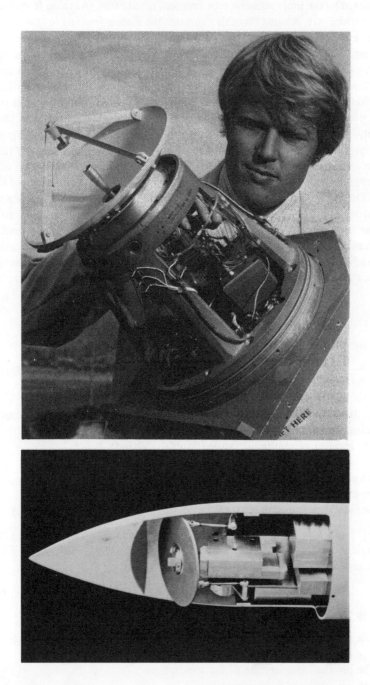

Figure 8.4   *Millimetre wave seekers (courtesy of Royal Military College of Science)*

be detected by the most sensitive receivers, or radiometers. (At radio frequencies, where $ch/\lambda kT \ll 1$, h being Planck's constant, the Planck formula for the spectral radiant emittance of a black body becomes $(2kT/\lambda^2)$ W m^{-2} Hz^{-1} Sr^{-1}.) In fact, targets are seen against a background, and the radiometer detects the target by detecting the difference in radiation between the two. Radiometers are no more than conventional receivers but with the best available noise performance and with antenna side-lobes as low as possible. The most straightforward is the 'total power receiver', which integrates the power, signal and noise. Unfortunately, variations in receiver gain or noise performance change the background noise level, and this tends to mask any change due to the presence of a target. Most elaborations of this receiver seek to provide continuous calibration of the background, thereby providing a known reference level. The best known of these is the Dicke receiver which compares the power received by the antenna with that generated by a thermal load at a known temperature. The input to the receiver is multiplexed in time between the antenna and the reference load at a sufficiently high rate, typically several 100 Hz, to cater for short-term variations; the disadvantage is that half the antenna signal power is lost. The detection range R for a target can be calculated by the radiometer range equation which is

$$\eta \times \frac{A_T A_R}{\lambda^2 R^2} \times \frac{\Delta T_T}{L} = \frac{CT_s S{:}N_{min}}{\sqrt{Bt_i}} \qquad (8.2)$$

for a target which does not fill the antenna beam. In this equation

A_T is the radiometric area of the target (closely related to the physical area)
A_R is the antenna aperture area
η is the aperture efficiency
ΔT_T is the radiometric temperature difference between target and background
L is the atmospheric transmission loss
T_s is the receiver system temperature (double sideband)
$S{:}N_{min}$ is the signal:noise ratio for the required probability of detection
B is the pre-detector (IF) bandwidth of the receiver
t_i is the integration time (reciprocal of the post-detector bandwidth), and
C is the radiometer constant (1 for total power receiver, 2 for Dicke receiver).

The left-hand side of equation 8.2 represents the target power, and the right-hand side represents the receiver sensitivity.

The temperature differential ΔT_T (ref. 12) is given by

$$\Delta T_T = E_T T_T + (1 - E_T)E_a T_a - E_e T_e - (1 - E_e)E_a T_a \qquad (8.3)$$

where in this equation

T_e is the physical temperature of the terrestrial background
E_e is the emissivity of the background
T_a is the physical temperature of the atmosphere
E_a is the emissivity of the atmosphere

T_T is the physical temperature of the target, and
E_T is the emissivity of the target.

Let us estimate the detection range of a tank.
T_e can be taken as 290K, and E_e for agricultural land is high, say 0.95.
T_a can also be taken as 290K, and E_a is low in an atmospheric attenuation window and in clear weather, being about 0.2 at 94 GHz.
T_T is of the order of 290K, though it may be hotter if the vehicle's engine is running; metal surfaces are reflective, hence E_T is low, say 0.1.

Hence from equation 8.3

$$\Delta T_T = 290 \; [0.1 + (0.9 \times 0.2) - 0.95 - (0.05 \times 0.2)] = -200K$$

Note that the target appears much cooler than the background.
For the other quantities assume

Antenna diameter = 80 mm and aperture

efficiency = 0.6, giving A_r as	$5 \times 10^{-3} \text{ m}^2$
A_T	10 m^2
λ	3.2 mm
C for total power receiver	1

$T_s = T_R = T_a \times E_a$, where T_R is the double sideband noise temperature of the receiver. At 94 GHz a reasonable value is 1440K, giving

T_s	1500 K

B is as wide as possible; at an IF of 4 GHz, using a FET as amplifier

B	1 GHz
t_i, which depends upon the servo system time constant	0.1 s
$S{:}N_{min}$ is quoted[13] for a probability of detection of 0.9 as	4:1

Substituting all these values in equation 8.2 gives

$$R^2 L = \frac{0.6 \times 10 \times 5 \times 10^{-3} \times 200 \times (10^9)^{\frac{1}{2}} \times 10^{-1}}{(3.2 \times 10^{-3})^2 \times 1 \times 1500 \times 4}$$

$$R \times L^{\frac{1}{2}} = 990 \text{ m}$$

At 94 GHz, $L = 0.4$ dB km^{-1} in clear weather, giving a corrected value

$$R = 950 \text{ m}$$

This example brings out the fact that radiometric detection is essentially short range, hence in battlefield applications it is best suited to TGSM.

8.5 Anti-radar missiles (ARM)[18]

As the name implies, the missile homes passively on to a source of enemy radar radiation. The missile may be launched from the air, or from the land or sea surface; it may acquire the target before launch, or during flight; it may be directed against a specific radar, or it may look for a target within a specific band. To date, ARM have been designed for homing on to static or slow-moving surface targets; it is, however, possible to envisage an ARM for homing on to a large airborne radar installation. The main criterion for the use of an ARM is that the target shall be of a sufficiently high value to warrant the expenditure of the weapon. Major air defence surveillance radars*, long and medium range missile guidance radars, and weapon location radars are examples of targets falling within this category. The ARM disables its target by destroying the antenna; for this purpose it usually carries a blast or fragmenting warhead.

Frequency coverage

Most large military radars occupy the frequency band of 1–18 GHz; ideally, an ARM seeker should be able to cover the whole of this band without change of components or assemblies. Unfortunately, it is not possible, at present, to cover more than a 10:1 frequency band, so that some restriction of frequency coverage is essential. Possible coverages are 1–10 GHz and 2–18 GHz. A suitable type of antenna for such wideband coverage is the cavity-backed flat spiral, engraved in metal strip on a thin insulating substrate[19]. The antenna array consists of four spirals arranged in a phase comparison monopulse configuration; this will be usually on a flat circular disc (figure 8.5), protected by a radome, but the antennas could be mounted conformally on the missile nose cone[20]. Assuming that the antennas are on a flat disc and that the spirals are nearly touching, the external radius of each spiral is $r = (2^{\frac{1}{2}} - 1)\,(d/2) = 0.414\,(d/2)$, where d is the disc diameter. The property of the spiral is that the longest usable wavelength, λ_L is equal to $4r$, equivalent to $0.828d$. For example, if $\lambda_L = 0.3$ m, then d must be 0.36 m. The gain of each antenna is about 2 over the 10:1 frequency band, with a half-power beamwidth of 90°; the sum of the gains is 8.

Guidance and control

Because the target is stationary, or nearly so, and because the missile is launched directly towards the target, the angle rates arising in proportional navigation are slow, thus permitting the use of a fixed antenna coaxial with the missile. The proportional navigation equation is (figure 8.6)

$$\dot{\psi}_m = \lambda \dot{\psi}_s \tag{8.4}$$

which is the same as that quoted in section 1.5, but equating $\dot{\psi}_f$ to $\dot{\psi}_m$. The missile measures $\dot{\psi}_m$ by gyroscope as usual, but it derives $\dot{\psi}_s$ by adding $\dot{\psi}_m$ to $\dot{\theta}_s$, the rate of change of sight-line direction, S, measured with respect to the missile axis M. This gives

*A British aircraft fired a Shrike ARM at a large Argentinian air defence surveillance radar during the recent war in the Falkland Islands; it damaged the antenna but the radar was out of action for only a few days.

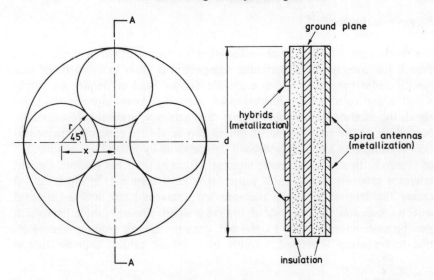

Figure 8.5 *Cluster of four flat spiral antennas*

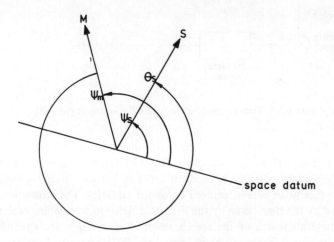

Figure 8.6 *Navigation geometry*

$$\dot{\psi}_m = \lambda(\dot{\psi}_m + \dot{\theta}_s)$$

$$\dot{\psi}_m = (\lambda/(1 - \lambda))\dot{\theta}_s \qquad (8.5)$$

This equation represents an unstable system unless $\lambda < 1$, but the method is adequate for homing in the circumstances postulated.

Receiver configuration[21]

The receiver must be able to cover the whole of the prescribed frequency band, though the coverage for a particular engagement is likely to be restricted to a specific radar band, or even to a specific narrow band of frequencies within which a particular radar is known to be operating. After acquiring the radar signal, the receiver must be able to gate the frequency, preferably to within the bandwidth of the transmission; it should also be able to cope with frequency-agile transmissions and other wideband transmissions by gating them in a number of channels. In short, the seeker requires a form of intercept receiver. Current intercept receivers use a tunable superheterodyne (figure 8.7) in which the LO causes the receiver to sweep the specified intercept band until a threshold detector indicates the presence of the radar signal. The acquisition bandwidth may be quite wide (10 MHz–1 GHz) but, once the signal is acquired, the resolution in frequency is refined, possibly by a second tunable superheterodyne

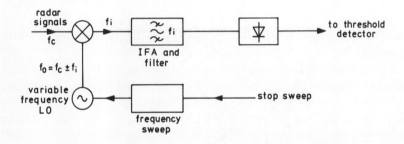

Figure 8.7   *Tunable superheterodyne radar intercept receiver*

amplifier. The first, wide-frequency range, LO may be tuned by an yttrium–iron–garnet (YIG) crystal or by a varactor; with YIG tuning it may be necessary to convert the signal frequency upwards in order to bring it within the coverage of the LO (typically an octave, centred on about 10 GHz). The acquisition time is determined by the time taken by the threshold detector to examine each acquisition bandwidth's worth of the search ambit. For example, for a search ambit covering the whole frequency band from 1 to 10 GHz with an acquisition bandwidth of 50 MHz, and with a threshold detector examination time of 10 ms, the acquisition time could extend up to 2 seconds. After initial acquisition, the frequency resolution is refined to, say, 1 MHz, adding a further 0.5 s to the acquisition time. This acquisition time is probably quite acceptable for a typical missile with a range of 25 km and a speed of 700 ms^{-1}, to give a flight time of 35 s.

More modern is the channelized receiver (figure 8.8) using a bank of narrow band filters covering a wide band of frequencies simultaneously. Each channel has its own threshold detector, and a detection logic circuit samples each channel in turn in a time equal to the reciprocal of the channel bandwidth. Since the

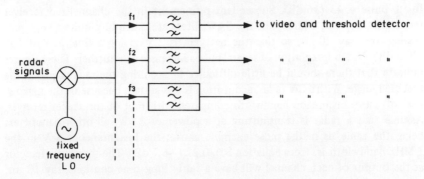

Figure 8.8   *Channelized radar intercept receiver*

bandwidth is at least a megahertz, acquisition is almost instantaneous. The narrow band filters are usually surface acoustic wave (SAW) devices; with current SAW technology the maximum frequency band which the channelized receiver can cover is about 250 MHz, and hence it is necessary to precede it with the older type of tunable superheterodyne receiver. Nevertheless, the acquisition time is much shorter. Since a 250 MHz channelized receiver can examine up to 250 channels simultaneously, it can cope with most broad-band or frequency-agile transmissions. The SAW compressive receiver offers similar performance, with slightly more elaboration but lower power consumption.

Acquisition

The seeker must have a very high probability of acquiring the target at maximum range; since there is no guarantee that the radar beam will be pointing at the missile, the seeker must rely on radiation in the radar antenna side-lobes. Calculations are made by using the equation for signal:noise in a one-way radio link, which is

$$S{:}N = \frac{P_t G_t G_m \lambda^2}{(4\pi)^2 R^2 (NF)(kT_0 B_n)} \qquad (8.6)$$

Assume a pulse radar, whose transmitted power, P_t, is 1 MW. Its one-way antenna gain is +25 dB with a general side-lobe level of −30 dB with respect to the main lobe, giving a gain, G_t, for the purpose of this calculation of −5 dB; the missile will be aimed at the radar, hence its gain, G_m, is the gain of the array, shown above to be 8. Assume λ is 0.3 m, corresponding to λ_L for a seeker coverage of 1–10 GHz, and assume that the range, R, is 25 km. Also assume that the receiver is a multichannel type with 250 channels, each of bandwidth 1 MHz, the whole being preceded by a tunable superheterodyne receiver of bandwidth, B_n, 250 MHz. For initial acquisition in the wide band receiver, a reasonable value of noise figure, NF, is 20 dB. Substituting these values in equation 8.6 yields $S{:}N$

for 1 pulse = 43 (16 dB). Subsequent refinement in the channelized receiver reduces B_n to, say, 1 MHz, but the SAW filters etc. introduce quite a large processing loss, say 20 dB, so that the net gain is 2,5, giving a final $S:N$ of 108 (20.3 dB). Video integration of several pulses increases $S:N$ further. These figures indicate that there should be little difficulty in acquiring the radar signal at the specified range. With a CW radar, acquisition is less certain because of the narrowness of the transmission bandwidth compared with that of the seeker receiver. Assume that a radar is transmitting at a power of 2 kW, all other parameters being the same as in the pulse example above; the instantaneous $S:N$ in the 1 MHz bandwidth is, from equation 8.6, 0.22 (−6.7 dB). The threshold detector at the output of each channel will have a fairly long time constant, say 10 ms, giving post-video detector signal integration; the integration improvement factor is $(10^6 \times 0.01)^{\frac{1}{2}} = 100$, raising $S:N$ to 22, which should be adequate. If the radar transmission is wide band, the signal will appear in all or several channels simultaneously; provided that the signal bandwidth is no wider than that covered by the channelized receiver, which it is unlikely to be, there is no loss of detectability.

Tracking precision

A direct hit on the antenna is desirable, calling for high precision of homing. Take as an example an antenna of size 5 m × 5 m; for a high probability of impact the standard deviation of the random homing error should not exceed 1/6th of 5 m = 0.83 m in each coordinate. Assume a time constant of 0.5 s for missile manoeuvre and a missile speed of 700 ms^{-1}; the corresponding permissible standard deviation in missile heading is 0.83/350 = 2.4 mrad. This will include instrumental errors, aerodynamic effects, and error due to receiver noise. The effect of receiver noise on the tracking precision of various seeker antenna systems is studied in section 4.6. The fixed antenna system is comparable to the interferometer with auxiliary sum signal antenna; in this case, however, the four interferometer antennas themselves provide the sum signal. The equation for the standard deviation due to receiver noise, σ_θ, is the same as equation 4.9 with the identification G_I = gain of each individual antenna and G_D = gain of the array of four antennas, giving $G_D/G_I = 4$; the interferometer spacing (d in equation 4.9) = $2x$ (see figure 8.5), hence

$$\sigma_\theta = 4^{\frac{1}{2}} \times \frac{\lambda}{2\pi x} \times \frac{\sec \theta_s}{[2(S:N)(BT)]^{\frac{1}{2}}} \qquad (8.7)$$

Provided that the signal is strong enough to actuate the AGC, $S:N$ is that in the pre-video detector bandwidth, B; T is the missile servo time constant in pitch or yaw. Note that since $S:N \propto \lambda^2$ (equation 8.6), σ_θ is nominally independent of λ. The missile should be heading directly at the radar, hence $\sec \theta_s$ can be taken as 1. Choose $\lambda = \lambda_L$, hence $\lambda = 0.828d$ (from the section on Frequency coverage). From figure 8.5, $x = 2^{\frac{1}{2}}r = 0.586d$. In a pulse radar, BT = the number of pulses, n, integrated in time T. From the example of the pulse radar in the section on

Acquisition, $S{:}N$ for 1 pulse on acquisition is 108; assuming a radar PRF of 1 kHz and $T = 0.5$ s gives $n = 500$; substituting this and other quantities already specified in equation 8.7 gives $\sigma_\theta = 1.37$ mrad at acquisition. As the range shortens $S{:}N$ increases, with a corresponding decrease in σ_θ, but only within the limit of saturation of the receiver. Since the signal is already strong, 1.37 mrad is a fair estimate of σ_θ throughout flight. By comparison with the total permissible standard deviation of 2.4 mrad, this figure is probably acceptable.

Counter-measures

The implementation of any counter-measure depends upon the ability of the radar under attack, or of an adjacent radar, to detect the ARM in good time. Consider the ability of the pulse surveillance radar, quoted in the previous examples in this section, to detect the ARM. The maximum detection range, $R_{\max}$, is given by

$$R^4_{\max} = \frac{P_t G_t{}^2 \sigma \lambda^2}{(4\pi)^3 \, (NF) \, (S{:}N_{\min}) \, (kT_0 B_n)} \tag{8.8}$$

where the symbols have their usual meanings. Assume $P_t = 1$ MW, $G_t = 25$ dB, σ of the ARM (nose on) $= 0.03$ m^2, $\lambda = 0.3$ m, radar receiver $NF = 10$ dB, radar $S{:}N_{\min} = +10$ dB, and $B_n = 1$ MHz. Substituting these values in equation 8.8 yields $R_{\max} = 24$ km; comparing this with 25 km nominal range at which the ARM can easily detect the radar confirms that the advantage lies with the ARM. If the missile speed is 700 m s^{-1}, the reaction time available for the deployment of counter-measures is less than 29 seconds. The radar can take the following measures. It can switch off, either on detecting a possible ARM, or at intervals, the purpose of switching off at intervals being to deprive any ARM of guidance information for long enough to cause it to lose the target. It can change frequency or, preferably, change frequency band; to which the ARM can reply only by initiating a fresh search. A nearby decoy transmitter can be switched on[22]; ideally, the radar transmitter and the decoy should 'blink', confusing the ARM and causing it to fly harmlessly between the two; the appropriate counter-measure is for the ARM to look for modulations peculiar to the radar transmitter and which may not exist on the cruder decoy transmissions.

8.6 ECM[23]

Chapter 6 introduced this subject prior to describing it relevant to air defence homers. The same general considerations apply in radar homing on to surface targets, with the proviso that most seekers are non-coherent pulse, thus affecting the choice of effective ECM. Until recently, homing on to surface targets was confined to ships at sea, consequently the main anti-missile ECM development effort has been in the context of naval warfare. There is now an upsurge of interest in radar homing on to large battlefield targets such as armoured fighting vehicles (AFV) using millimetre wavelengths, and corresponding developments in ECM are sure to follow. This section, therefore, concentrates on naval applications, but discusses briefly what may be feasible on the battlefield.

Noise jamming (confusion ECM)

Each ship of a force carries a complete electronic warfare outfit consisting of both electronic support measures (ESM) and ECM, so that each ship can not only protect itself but can also help to protect other ships of the formation. The purpose of noise jamming is to deprive the attacking missile of range information, leading it to attempt to home on to a source of radiation which, it is hoped, will be out of propulsion range. The following numerical example illustrates the effectiveness of this type of ECM. Assume that the parameters of the seeker of an active terminal homing sea-skimmer are: transmitter power P_I = 10 kW, antenna gain G_I = 27 dB with a general side-lobe level of −20 dB with respect to the main lobe, receiver bandwidth B = 10 MHz, and $S{:}N_{min}$ = +6 dB. Also assume that the radar cross-section of the target σ is 100 m^2 (a small target, or a large target at an unfavourable aspect) − the homing phase is 12 km. An escort vessel standing off at a mean range from the missile R_{JM} of 20 km generates a noise jamming ERP density P_J of 1 kW per MHz. From equations 6.12 and 6.13 of section 6.11, the jammer denies range information to the seeker until the missile is within a range R from the target given by

$$R^4 = \frac{P_I G_I}{F P_J B} \times \frac{\sigma}{4\pi} \times \frac{R_{JM}^2}{(S{:}N)_{min}} \qquad (8.9)$$

Since the antenna is likely to receive the jamming in a side-lobe, F = 0.01. Inserting the numerical data in this equation gives R = 4.5 km, which is significantly shorter than 12 km; indeed, since the missile is probably closing on the source of ECM as well as on its designated target, R is likely to be even less. An appropriate ECCM would be examination of the angle error signal in two adjacent narrow band filters after demodulation, as described in section 6.11, in order to reject any signal which is off-axis.

Chaff[24]

This is a potent ECM because the seeker is unlikely to be able to distinguish between the chaff and a genuine target on the basis of relative velocity. A ship dispenses chaff by means of a rocket or a mortar so that it forms a cloud at a safe distance from the ship, say up to 2 km. According to Boyd *et al.*[25], 0.136 kg of mixed chaff covering three octaves can generate a cloud of maximum echoing area 47 m^2, hence a rocket package of 7 kg should generate a cloud of about

2000 m^2 *. This is enough to create a distraction from quite a large vessel. A cloud of aluminium foil chaff forms quite quickly, attaining half its maximum in a second or two and attaining its maximum in about 10 seconds; it decays again to half its maximum in about 60 s[26]. An effective way of using chaff in conjunction with active ECM is to deny range information to a missile seeker by noise jamming until the missile is close in; in the meantime, clouds of chaff will have been fired to present more attractive targets than the ship under attack. Should this fail initially to decoy the missile, the target uses a range gate stealer to unlock the seeker range gate from the genuine echo and to dump it on to a nearby cloud of chaff. A possible ECCM is to distinguish chaff from a genuine target by its broader frequency spectrum and by its random drifting motion. Such an examination takes time, however — at least a few seconds; if the seeker is prevented from acquiring a target until it is within 4.5 km, and if the missile flies at a high subsonic velocity, say 300 m s^{-1}, only 15 seconds are available.

ECM against battlefield homing missiles[27]

Effective noise jamming is likely to be difficult to implement. The millimetre bands are wide and the terminal phase of battlefield homing is short, allowing little time for ESM; this will force noise jammers to dilute their effort in an endeavour to cover the whole band. No convenient high-power mm wave sources are available at present, hence it will be difficult to achieve an adequate power density. Chaff fired from projectors on AFVs is a more likely ECM; the radar cross-section of an AFV lies between about 10 m^2 and 100 m^2, depending on aspect, hence quite small packages of chaff can provide adequate decoying. An even simpler defence is to spread sheets of metal on the ground to act as decoys, though this is only practicable if the AFV formation being protected is stationary. A simple rotating reflector on a vehicle could upset conical scan. Battlefield homing seekers, especially those in TGSMs, are simple and cheap, with no place for elaborate ECCM facilities; coherent radar is easy enough to implement and would give a measure of discrimination against static decoys and chaff when engaging moving targets. However it is likely that the most reliance will be placed upon sufficiently large numbers of missiles with frequency diversity to ensure that some get through.

References

1. Nathanson, F. E., *Radar Design Principles*, McGraw-Hill, 1969, Fig. 7.1.
2. Skolnik, M. I., *Introduction to Radar Systems*, 2nd edn, McGraw-Hill, 1980, Eqn 2.38.

---

*Scaling is, however, likely to be less than proportional to mass of package because of mutual coupling between dipoles in a large dense cloud.

3. Skolnik, M. I., *Introduction to Radar Systems*, 2nd edn, McGraw-Hill, 1980, Fig. 2.21.
4. Nathanson, F. E., *Radar Design Principles*, McGraw-Hill, 1969, Table 7.2.
5. Nathanson, F. E., *Radar Design Principles*, McGraw-Hill, 1969, Table 7.7.
6. Ramsay, D. A., 'Missile radar guidance', *Symposium on Trends in Missile Guidance*, Royal Aeronautical Society, 14 January 1981.
7. Seashore, C. R. and Singh, D. R., 'MM-wave component trade-offs for tactical systems', *Microwave Journal*, June 1982, p. 41.
8. Johnston, S. L., 'Millimetre-wave anti-tank guided missiles', *Conference Proc. Military Microwaves MM82*, pp. 133–40.
9. Kuno, H. J., 'Are millimetre-wave systems affordable now?', *Microwave Journal*, June 1982, p. 16.
10. Trebits, R. N., *et al.*, 'mm-wave reflectivity of land and sea', *Microwave Journal*, August 1978, p. 49.
11. Brookner, E., 'Present and future trends in synthetic aperture radar systems and techniques', in Brookner, E. (ed.), *Radar Technology*, Artech House, 1977, Chapter 18.
12. Tipping, D. E. J., 'Systems requirement for the precision guided munition', in IEE Electronics Division (Professional Group E15), *Colloquium on millimetre-wave radar and radiometer sensors, Digest*, 1980/23.
13. Henderson, A. and James, J. R., 'A survey of millimetre wavelengths planar antenna arrays for military applications', *Radio and Electronic Engineer*, Vol. 52, no. 12, pp. 543–50.
14. McGillem, C. D. and Seling, T. V., 'Influence of system parameters on airborne microwave radiometer design', *IEEE Trans. MIL.*, October 1963, pp. 296–302.
15. Schuchardt, J. M., *et al.*, 'The coming of mm-wave forward looking imaging radiometers', *Microwave Journal*, June 1981, p. 45ff.
16. Seashore, C. R., *et al.*, 'mm-wave radar and radiometer sensors for guidance systems', *Microwave Journal*, August 1979, pp. 47–51, 58.
17. IEE Electronics Division (Professional Group E15), *Colloquium on millimetre-wave radar and radiometer sensors, Digest*, 1980/23.
18. Boyle, D., 'Anti-radar missiles', *Interavia*, No. 11/1982, pp. 1194–5.
19. Stutzmann, W. L. and Thiele, G. A., *Antenna Theory and Design*, Wiley, 1981, Section 6.4.
20. Godfrey, M. F., 'Homing Heads for Guided Missiles', *Conference Proc. Military Microwaves MM82*, late paper.
21. Grant, P. M. and Collins, J. H., 'Introduction to Electronic Warfare', *IEE Proc.*, Vol. 129, Part F, No. 3, June 1982, pp. 113–130.
22. Ellis, A. T., Marc, S. and Dulieu, M., 'New Decoy Systems for Ships', *Conference Proc. Military Microwaves MM82*, late paper.
23. Grant, P. M. and Collins, J. H., *Conference Proc. Military Microwaves MM82*.
24. Butters, B. C. F., 'Chaff', *IEE Proc.*, Vol. 129, Part F, No. 3, June 1982, pp. 197–201.

25. Boyd, A. J., *et al., Electronic Countermeasures*, Peninsular Publications, 1978.
26. Butters, B. C. F., 'Chaff', *IEE Proc.*, Vol. 129, Part F, No. 3, June 1982, Fig. 4a.
27. Lorber, A. K., 'Active defence against precision guided munitions for the future battlefield', *Military Technology*, 5/1985, pp. 77-82.

9

Future Developments

9.1 Introduction

Homing guidance for tactical missiles has come a long way since it was first considered seriously, just after the end of the Second World War, as a logical extension of radar. A review of more recent developments which have led to the present state of the science of missile guidance appears in ref. 1. This brings out the strong influence of the growth of digital techniques and large scale integration, of microwave microcircuits and array antennas, and of microwave solid state sources. In addition, with the growth of millimetre-wave technology the anti-vehicle homer has become a practical proposition.

It is considered that these four areas of technology, together with improvements in radome material and manufacture, will continue to have the most decisive effect on developments in radar seekers.

The aim of this chapter is to discuss each of these five points in the context of operational significance.

9.2 Digital signal processing

Digital technology is reaching the stage of very large scale integration and of very high speed integrated circuits; a sample of performance goals for 1985 includes 450 000 transistors on a chip, operating at a clock rate of 18 MHz, a density of 50 000 gates cm^{-2} with a clock rate of 10 MHz, leading eventually to 100 000 gates cm^{-2} at 100 MHz clock rate, and a 40 MCOPS processor on two 6 inch by 8 inch printed circuit boards[2]. It can be inferred that it is already feasible to undertake all the processing of signals in those stages of the receiver that follow the main IF amplifiers of a pulse-Doppler seeker by digital means for present-day airborne targets. Indeed, this present state of the technology is likely to be adequate for meeting new seeker requirements for some time to come.

Modern SAM and AAM systems are already taking advantage of these techniques to acquire the versatility and speedy reaction demanded by the modern dense airborne target environment. It is probable that older but still useful systems will be updated by replacement of the existing analogue processing.

Digital signal processing is likely to be a major feature of seekers for ARM. Although frequency analysis still has to be performed by analogue techniques, some of which are quite elegant, digital methods are very valuable for the rapid analysis of signal characteristics, for the rejection of interference and ECM, and for managing the acquisition and tracking process.

Signal processing in anti-ship missiles tends to be less demanding because the basic homing process requires information in angle and range only, and because targets are slow and usually fairly well dispersed. Nevertheless there are likely to be calls for echo analysis for identifying targets, for rejecting chaff and other ECM, and for rejecting clutter, particularly in the vicinity of land. These tasks are almost certain to be undertaken by digital means.

Battlefield seekers are in their infancy and are therefore still simple in concept. However, as more sophisticated methods of target assessment and recognition are demanded and as the ECM threat grows, there will be a need for a measure of signal processing. In view of the small size of this type of missile, the processing is certain to be digital.

9.3 Millimetre-wave technology

The present interest arises from the growing availability of solid state sources, particularly the IMPATT diode, and of low noise receiver devices[3,4,5]. There are, of course, limitations. Propagation characteristics limit the range in the troposphere and there is an upper limit to the power available from solid state sources, which falls rapidly with increase of frequency; a typical figure is 1 W CW at 100 GHz. The power available from thermionic sources is much greater but the associated high voltage power supply and the weight of ancillaries makes them unattractive in the small size of missile envisaged. From these considerations the most likely applications of mm-wave technology are battlefield homing missiles, which are already in existence, and active terminal seekers for medium and high level SAM and AAM, possibly as supplementary to an existing method of homing. It is also possible that a mm-wave seeker could be used against very low level targets, as a better all-weather alternative to IR homing. Millimetre waves are unlikely to be employed in an anti-ship seeker, even at short range, because the clutter level[6] tends to offset any advantage of better resolution.

9.4 Array antennas

The flat plate array is already well established as a seeker antenna at centimetric wavelengths and developments at mm wavelengths are proceeding[7]. The prospect for fixed active antennas (phased arrays)[8] is less sure. The advantage of no moving parts tends to be outweighed by the complexity of the phase-shifting circuitry. Furthermore, no way has yet been devised of relating the beam axis directly to an inertial frame of reference; until then, the use of these arrays is limited to homing on stationary, or nearly stationary, targets.

Conformal arrays are especially attractive in that they enable the conventional radome to be dispensed with; unfortunately the problems associated with the geometry of the array are formidable and obscuration of elements whenever the beam is at a squinting angle is unavoidable. Although interest is sustained, no practical example seems likely to go into service for some years.

9.5 Microwave solid state sources

The impact of solid state sources on mm-wave technology has been mentioned already. With microwaves the FET is already capable of providing a receiver whose noise figure is good enough for all foreseeable purposes. The incorporation of active components into microwave integrated circuits is a big step towards miniaturizing the seeker.

Development of the solid state microwave power sources, particularly the IMPATT diode and the transistor, continues in an effort to achieve higher power and better noise performance. There is, however, a theoretical upper limit of power, which is about 100 W CW at 10 GHz; this is probably adequate for most active terminal seeker applications, though it must be remembered that this same average power cannot be achieved at low duty cycles. Poor noise performance is a disadvantage compared with thermionic devices; the sideband noise is 20–30 dB worse, which would create difficulties in a Doppler seeker. Every effort will doubtless be made to achieve a more compact, more robust and lighter seeker by employing a solid state transmitter, but it is likely to be some time before this becomes a general possibility.

9.6 Radome technology

Much valuable development has taken place already and the new materials and methods of manufacture mentioned in chapter 4 are likely to meet new radar requirements for some years to come. Nevertheless, new materials are always being developed or being tried and some are certain to be superior in one or more respects. For example, thermoplastics are now being studied and several materials show promise for high performance radomes[9]. This is a healthy state of affairs for, although radome technology does not drive guided missile seeker development, it is important that radomes are availabe as new requirements materialize.

9.7 Summary

Radar homing guidance is now at a mature stage of development, especially in its original role against airborne targets. In SAM and AAM there is likely to be a period of consolidation, with detail improvements stemming from advances in existing technology and with strong emphasis on resistance to ECM.

ARM are likely to proliferate, with cheaper and smaller missiles bringing more types of radar into the category of 'worthwhile' targets. The larger types of ARM will possess a greater measure of autonomy, based upon increased ability to select suitable targets and to reject decoys.

Second generation anti-ship, sea-skimming missiles are already entering into service; they are certain to be well equipped with ECCM and to contain means for distinguishing between suitable and unsuitable targets.

Homing against battlefield targets presents the greatest scope for innovation; to a great extent this will be dictated by whatever forms ECM and other counter-measures take. At the same time efforts will probably be made to lengthen range. Whatever the developments, this new method of attacking vehicles, whether armoured or not, is likely to have a profound effect.

References

1. Tiernan, K. E., 'Missile guidance responds to electronic technology', *Electronics for National Security, Conference Proceedings, 27–29 September 1983*, Interavia S.A., pp. 137–147.
2. Brookner, E., 'Developments in digital radar processing', in Oppenheim, B. V. (ed.), *Trends and Perspectives in Signal Processing*, Vol. 2, No. 1, January 1982, p. 7.2ff.
3. Malbon, R. M. and Osbrink, N. K., 'Small signal low noise active devices', *Microwave Journal*, February 1985, p. 121ff.
4. Gupta, C. and del Conte, J., 'Trends in mm wave mixer designs', *Microwave Journal*, April 1984, p. 83ff.
5. Seashore, C. R. and Singh, D. R., 'MM-wave component trade-offs for tactical systems', *Microwave Journal*, June 1982, p. 41ff.
6. Trebits, R. N., *et al.*, 'MM wave reflectivity of land and sea', *Microwave Journal*, August 1978, p. 49ff.
7. Henderson, A. and James, J. R., 'A survey of millimetric wavelength planar array antennas for military applications', *Radio and Electronic Engineer*, Vol. 52, No. 11/12, 1982, pp. 543–580.

8. Maurer, H. A., 'MMW seekers − scaled-down radars or new concepts?', *Electronics for National Security, Conference Proceedings, 27-29 September 1983*, Interavia S.A., pp. 165-168.

9. Hall, C. K., 'Radome materials, contemporary trends', *Military Microwaves 1984, Conference Proceedings*, pp. 367-387.

Appendix A:

Derivation of k for Phase Comparison Seeker

(see chapter 4)

Each half antenna constitutes a semi-circular planar array. Assuming that the distribution of elements is uniform, the phase centre of each half lies at the centroid, a distance $2d/3\pi$ from the centre O. The two halves constitute an interferometer pair, each of gain $G/2$ with spacing $d' = 4d/3\pi$.

from equation 4.1;
$$\Delta = G^{\frac{1}{2}} \sin\left(\frac{\pi d'}{\lambda} \sin\theta_T\right)$$

$$\left.\frac{d\Delta}{d\theta_T}\right|_{\theta_T = 0} = G^{\frac{1}{2}} \times \pi d'/\lambda$$

also
$$\Sigma = G^{\frac{1}{2}}$$

$$\left.\frac{d(\Delta/\Sigma)}{d\theta_T}\right|_{\theta_T = 0} = \frac{\pi d'}{\lambda}$$

$$= 4d/3\lambda$$

$$k = \frac{4d}{\lambda}\,\theta_3 \qquad\qquad (A.1)$$

Derivation of k for interferometer with auxiliary antenna

from equation 4.3
$$\Delta = 2G_I \sin\left[\frac{\pi d}{\lambda}\cos\left(\frac{\theta_s + \theta_D}{2}\right)\sin\left(\frac{\theta_s - \theta_D}{2}\right)\right]$$

where G_I = interferometer antenna gain.

$$\frac{\mathrm{d}\Delta}{\mathrm{d}\theta_T}\bigg|_{\theta_T=0} = (2G_I)^{\frac{1}{2}} \times \frac{\pi d}{\lambda} \times \cos\theta_s, \text{ where } \theta_s - \theta_D = \theta_T$$

$\Sigma = G_D^{\frac{1}{2}}$, where G_D is the auxiliary antenna gain

$$\frac{\mathrm{d}(\Delta/\Sigma)}{\mathrm{d}\theta_T}\bigg|_{\theta_T=0} = \left(\frac{2G_I}{G_D}\right)^{\frac{1}{2}} \times \frac{\pi d}{\lambda} \times \cos\theta_s \qquad (A.2)$$

Appendix B:

Generation of False Target Signals in CW Homing Missile Receivers by Spurious Noise Modulation

(see chapter 6)

B.1 Introduction

The strongly vibrating environment within a missile can set up mechanical oscillations in critical electronic components and circuits, such as the receiver local oscillator and the receiver amplifier stages, causing unwanted amplitude and phase modulation of a carrier in the receiver channel. If any of the vibration frequency sidebands fall within the Doppler band of the system, and are of sufficient amplitude, they may be interpreted falsely as the Doppler frequencies of genuine target echoes. Vibration which is noise-like rather than at discrete frequencies degrades the signal:noise ratio. Noise and vibration modulations generated in the illuminator have similar effects.

Both amplitude (AM) and angle modulation (PM) occur. The AM is demodulated in the receiver in the same way as a genuine signal, and will be interpreted as such unless the amplitude is less than the minimum detectable signal amplitude. If the spurious AM is noise-like, then its power summed over the Doppler filter bandwidth must be less than the minimum detectable signal power.

PM is less obvious in its effects since the Doppler mixer rejects angle modulation. However, asymmetries and non-linearities in the receiver system can convert angle to amplitude modulation which can then pass the Doppler mixer. Two ways in which this can occur are described; both depend upon the presence of a strong carrier, such as that of the direct clutter or of the ground spike. In a late narrow-banding receiver, these carriers bear the modulations through the main IF amplifier to the Doppler mixer where, if there has been conversion from PM to AM, the modulations may appear as signals within the Doppler band. An early narrow-banding receiver is less subject to these effects, as the strong carriers concerned are rejected at the beginning of the chain of amplification. Nevertheless, stringent control of the generation of spurious modulation, both within the receiver and the illuminator, remains of paramount importance in a Doppler system.

B.2 Asymmetry of the signal channel IF amplifier frequency response

Figure B.1 shows a normalized response curve exhibiting a simple form of asymmetry — the gain within the pass band varies linearly with sideband frequency, f_m. Let the normalized gain be represented by $(1 + f_m q)$ where q is the slope of the gain characteristic. Consider vibration at a single frequency f_m Hz, then the LO frequency in Hz is $(f_0 + \Delta f \cos 2\pi f_m t)$, where Δf is the frequency

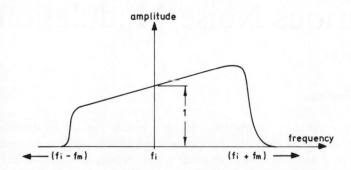

Figure B.1   *Asymmetry of IF amplifier amplitude response*

deviation. The FM is transferred in the first mixer to the zero Doppler frequency direct clutter to give a clutter IF voltage of frequency $(f_i + \Delta f \cos 2\pi f_m t)$ Hz. In practice, Δf will be of the order of a few hundred Hz, whereas vibration frequencies of interest, those within the Doppler band, will be of the order of kHz. This means that the modulation index β $(= \Delta f/f_m)$ is $\ll 1$, and this voltage can be represented as shown in the phasor diagram, figure B.2a, by a carrier of magnitude unity and two first order sidebands of magnitude $\beta/2$. Because of the asymmetry, the two sidebands are amplified unequally and the phasor diagram of the voltage emerging from the IF amplifier is as shown at b. Now the sidebands are shown at c to be equivalent to the original FM sidebands of modulation index β plus a pair of AM sidebands of depth $\beta f_m q$ $(= q \, \Delta f)$. If the ratio of clutter power to minimum detectable signal power is 60 dB, then after the second mixer the ratio

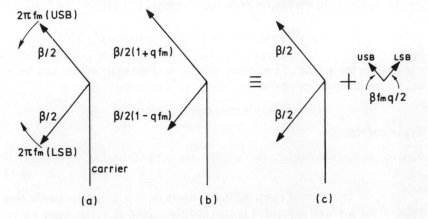

Figure B.2   *Phase modulation sidebands*

of clutter amplitude (V_c), which is of course at DC and is easily removed, to minimum signal amplitude (V_{min}) is 10^3. The second mixer also demodulates the vibration AM sidebands to generate a video voltage of amplitude $q\Delta fV_c$ at a frequency f_m. If f_m lies within the Doppler band, this can be interpreted as a target signal unless $q\,\Delta f < 10^{-3}$. For example, say q = 5 per cent per 100 kHz, then the maximum permissible value of Δf is given by

$$0.05 \times 10^{-5}\,\Delta f < 10^{-3}$$

$$\Delta f < 2\text{ kHz}$$

B.3 Differential time delay between reference and signal IF channels

As described in section B.2, the phase modulation of the LO frequency is transferred to the zero Doppler frequency clutter in the signal channel to give a phase modulated clutter voltage of *phase* ($\omega_i t + \beta \sin \omega_m t$) rad*. It is also transferred to the reference voltage in the reference first mixer to give a phase modulated reference signal of exactly the same phase. If there is a differential time delay of τ seconds in the passage of the voltage through the two amplifiers, the relative phases with which they emerge will be

Signal voltage $\omega_i t + \beta \sin \omega_m t$

Reference voltage $\omega_i(t - \tau) + \beta \sin \omega_m (t - \tau)$

The second mixer is a balanced mixer which generates an output voltage of magnitude proportional to the clutter voltage amplitude and to the cosine of the phase difference between the two signals; thus

*In this section, ω is used instead of f for the sake of brevity.

$$V_c \cos\{\omega_i\,\tau + \beta\,(\sin\omega_m t - \sin\omega_m\,(t-\tau))\}$$

$$= V_c \cos\left\{\omega_i\,\tau + 2\,\beta\,\sin\left(\frac{\omega_m\tau}{2}\right) \times \cos\omega_m\,(t-\tau/2)\right\}$$

τ is small (a few tenths of a microsecond), f_m is of the order of kHz, and hence $\omega_m\tau$ is small; remembering that $\beta\omega_m = \Delta\omega$, the expression reduces to

$$V_c \cos\{\omega_i\tau + \tau\Delta\omega\cos(\omega_m\,(t-\tau/2))\}$$

Expanding this gives

$$V_c\,[\cos\omega_i\tau\,\cos\{\tau\Delta\omega\,\cos(\omega_m\,(t-\tau/2))\} - \sin\omega_i\tau\,\sin\{\tau\Delta\omega\,\cos(\omega_m(t-\tau/2))\}]$$
$$\text{(B.1)}$$

There are two extreme cases; the first is where $\omega_i\tau = \pi/2$. In this case the first term of the previous expression is zero and the second term represents a sinusoidal voltage of frequency f_m; remembering that $\tau\Delta\omega \ll 1$ gives the amplitude of this voltage as $V_c\,\tau\,\Delta\omega$. It will be interpreted as a target signal unless its amplitude is less than V_{min}. Using the same ratio, of $V_c : V_{min} = 10^3$, as in section B.2, and taking $\tau = 0.5\ \mu$s, gives the maximum permissible value of Δf as

$$5 \times 10^{-7}\,\Delta\omega < 10^{-3}$$

$$\Delta\omega < 2 \times 10^3$$

$$\Delta f < 320\ \text{Hz}$$

The second extreme is where $\omega_i\tau = \pi$; the second term is now zero and the first term can be expanded to

$$V_c\left[1 - \left(\frac{\tau\Delta\omega}{2}\right)^2 \times \cos^2\,(\omega_m\,(t-\tau/2))\right] \qquad \text{(B.2)}$$

This represents DC plus a voltage at frequency $2f_m$. However, the amplitude of this voltage contains the term $(\tau\Delta\omega)^2$ and, since $(\tau\Delta\omega) \ll 1$, this voltage is much smaller than that of the first case. So it is the first case which determines the permissible frequency deviation of the LO.

B.4 Bibliography

A useful account of stability requirements in Doppler radar appears in Leeson, D. B. and Johnson, G. F., 'Short term stability for Doppler radar', Paper V-2 in Barton, D. K. (ed.), *Radars, Vol. 7, CW and Doppler Radar*, Artech House, 1978.

Appendix C:

Fast Fourier Transform
(see chapter 7)

The Discrete Fourier Transform of the continuous periodic function $f(t)$ of period T is given by equation 7.11 as

$$F_k = \frac{2}{N} \sum_{r=0}^{N-1} f(r)\exp(-j2\pi kr/N)$$

For the sake of brevity, write $\exp(-j2\pi kr/N)$ as W_{kr}, where the suffix is the product $k \times r$. The computation for all values of F is expressed by the matrix equation

$$
\begin{bmatrix}
F_0 \\
F_1 \\
F_2 \\
\cdot \\
F_{N-1}
\end{bmatrix}
=
\begin{bmatrix}
W_0 & W_0 & W_0 & \cdots & W_0 \\
W_0 & W_1 & W_2 & \cdots & W_{N-1} \\
W_0 & W_2 & W_4 & \cdots & W_{2(N-1)} \\
\cdot & \cdot & \cdot & \cdots & \cdot \\
W_0 & W_{N-1} & W_{2(N-1)} & \cdots & W_{(N-1)^2}
\end{bmatrix}
\begin{bmatrix}
f_0 \\
f_1 \\
f_2 \\
\cdot \\
f_{N-1}
\end{bmatrix}
\tag{C.1}
$$

showing that it needs N^2 complex multiplications to calculate all the values of F. The purpose of the FFT is to reduce this number significantly.

Take as example $N = 4$, so that

$$
\begin{bmatrix}
F_0 \\
F_1 \\
F_2 \\
F_3
\end{bmatrix}
=
\begin{bmatrix}
W_0 & W_0 & W_0 & W_0 \\
W_0 & W_1 & W_2 & W_3 \\
W_0 & W_2 & W_4 & W_6 \\
W_0 & W_3 & W_6 & W_9
\end{bmatrix}
\begin{bmatrix}
f_0 \\
f_1 \\
f_2 \\
f_3
\end{bmatrix}
\tag{C.2}
$$

Now W_{kr} repeats itself every time $2kr/N$ exceeds 2 – that is, kr is 'modulo N'. In this example $W_4 = W_0$, $W_6 = W_2$, $W_9 = W_1$, and $W_0 = 1$. Also $W_3 = -W_1$ and $W_2 = -W_0 = -1$, so that the equation becomes

$$
\begin{bmatrix} F_0 \\ F_1 \\ F_2 \\ F_3 \end{bmatrix}
\begin{bmatrix} 1 & 1 & 1 & 1 \\ 1 & W_1 & -1 & -W_1 \\ 1 & -1 & 1 & -1 \\ 1 & -W_1 & -1 & -W_1 \end{bmatrix}
\begin{bmatrix} f_0 \\ f_1 \\ f_2 \\ f_3 \end{bmatrix}
\tag{C.3}
$$

The different multiplications are now only $|1| \times f_0, f_1$ etc. and $|W_1| \times f_0, f_1$ etc., making eight in all, compared with sixteen of the original equation. By extension of the principle, it follows that the total number of complex multiplications is reduced from N^2 to $N \times \log_2(N)$. There are several ways of realising this FFT, one of the best known being the Cooley–Tukey algorithm. By this algorithm a start is made on calculating an F before the calculation for the previous F is complete, so that the calculations for several values of F can be in the 'pipeline'. This reduces the computation time by a further factor of up to 2.

Index

acceleration lag   67, 68, 83
accelerometer, missile   83
acquisition
   airborne target   83–6
   by ARM   121–2
   by sea-skimming missile   110–11
   in angle   83, 86, 110
   in Doppler frequency (velocity)
         70–2, 75–6, 84, 86, 93, 100
   in range   100, 110
   in terminal phase   84–6, 110
active homing   *see* homing
air-to-air missile (AAM)   47, 79, 96
air-to-surface missile (ASM)   14, 110
altimeter   6, 107
ambiguity
   angle   46
   Doppler frequency (velocity)   28,
         96, 97, 113
      resolution of   101–3
   range (elapsed time)   28, 97, 113
      resolution of   101–3
amplifiers
   baseband   72
   IF   63, 74
   low-noise   36
analogue filters   100
antenna
   amplitude comparison   38–41
   aperture area   32
   beamwidth (half-power)   7, 34

Cassegrain feed   40
cavity-backed flat spiral   118
conformal array   47, 130
end-fire   44
fixed   44, 47, 114, 118
flat plate array   42, 48, 130
gain   30, 31
phase comparison   48
phased array   11, 47, 49
rear reference   11, 78
   on missile   12, 64, 78
reflector type   7, 114
servo system   5
side-lobes   47, 57, 92, 116
squint angle   7
anti-radar missile (ARM)   15,
      117–23
artillery shell   1, 111
attenuation
   atmospheric   32, 34, 111
   motor flame   80
   radome   52
   rain   34
      effect of climate   35
automatic frequency control (AFC)
      10, 23, 64, 73
automatic gain control (AGC)   30, 63,
      64, 68, 79
autonomous, autonomy   3, 11, 12,
      33
autopilot   2

141